ECONOMIC ISSUES, PROBLEMS AND PERSPECTIVES

INTELLECTUAL PROPERTY

INDUSTRY CONTRIBUTIONS AND TRADE PROTECTION POLICY

Economic Issues, Problems and Perspectives

Additional books in this series can be found on Nova's website under the Series tab.

Additional E-books in this series can be found on Nova's website under the E-book tab.

Trade Issues, Policies and Laws

Additional books in this series can be found on Nova's website under the Series tab.

Additional E-books in this series can be found on Nova's website under the E-book tab.

ECONOMIC ISSUES, PROBLEMS AND PERSPECTIVES

INTELLECTUAL PROPERTY

INDUSTRY CONTRIBUTIONS AND TRADE PROTECTION POLICY

NICOLAS VALLÉE
EDITOR

nova
publishers
New York

For permission to use material from this book please contact us:
Telephone 631-231-7269; Fax 631-231-8175
Web Site: http://www.novapublishers.com

Additional color graphics may be available in the e-book version of this book.

Library of Congress Cataloging-in-Publication Data

ISBN: 978-1-62417-313-4

Published by Nova Science Publishers, Inc. † New York

CONTENTS

PREFACE

This book attempts to identify the first-order players that are growing IP in the U.S. economy and protecting their innovations through patents, trademarks, or copyrights. These IP-intensive industries support tens of millions of jobs and contribute several trillion dollars to our GDP, and represent the leading edge of our economy that is built on the ingenuity of the American people.

Chapter 1 - As President Obama has said, if we are to win the future and be successful in an increasingly competitive international market, the United States of America must innovate. Innovation, the process through which new ideas are generated and put into commercial practice, is a key force behind U.S. economic growth and national competitiveness. Likewise, U.S. companies' use of trademarks to distinguish their goods and services from those of competitors enables firms to capture market share and further strengthen our economy. The dynamics of a globally connected market mean that the United States will need to develop the brightest minds with the most advanced training to make the best products.

The Obama Administration's determination to promote innovation and protect intellectual property (IP) rights will harness the inherent drive and ingenuity of the American people in meeting that goal.

Innovation protected by IP rights is key to creating new jobs and growing exports. Innovation has a positive pervasive effect on the entire economy, and its benefits flow both upstream and downstream to every sector of the U.S. economy. Intellectual property is not just the final product of workers and companies—every job in some way, produces, supplies, consumes, or relies on innovation, creativity, and commercial distinctiveness. Protecting our ideas and IP promotes innovative, open, and competitive markets, and helps ensure that the U.S. private sector remains America's innovation engine.

There is a broad range of industries that benefit from IP, both directly and indirectly, and for every innovation in a given industry, generally there are corresponding economic opportunities for other industries to bring advances to the public. Examples of these complementary industries include the computer manufacturer that uses inputs made by semiconductor firms to make the hardware that is needed to run applications made by software companies, the Internet company that generates the on-line applications to distribute copyrighted music, and the auto manufacturer that incorporates patented energy-saving engines in the cars it sells. In each of these cases, industries are supported by the complementary products and services of another industry, so each industry is in a position to benefit from the safeguard of the IP underpinning their business models. When companies are more confident that their ideas will be protected, they have the incentive to pursue advances that push efficiency forward, costs down, and employment up.

Further evidence of this domino effect is seen as downstream businesses benefit from innovative products that lower their costs and improve their processes and finished articles. For example, a more precise machine could make manufacturing pharmaceuticals safer; a more reliable software program could improve military hardware while making it more affordable; or a clearer sound system could boost the quality of a recording studio's album. In each of these cases, and many more, the innovation feeds directly into a new finished article or service that has commercial value. The innovation may increase customer satisfaction through a higher quality product or service or decrease production costs. This additional value can help businesses protect earnings that can support their labor force.

Finally, in addition to the companies that produce, complement, and consume the innovation, logistical and supporting businesses are required to keep these companies up and running. These entities include marketing firms that inform consumers about innovations that can save time and money, packaging companies that ship advanced goods to retailers, and Internet service providers that create and maintain the communications pathways needed to compete in an increasingly on-line world.

This chapter by the Economics and Statistics Administration (ESA) and the U.S. Patent and Trademark Office (USPTO) of the U.S. Department of Commerce attempts to identify the first-order players that are growing IP in the U.S. economy and protecting their innovations through patents, trademarks, or copyrights. These IP-intensive industries support tens of millions of jobs and contribute several trillion dollars to our gross domestic product (GDP). This report not only estimates the contributions of these industries to our economy, but also gauges the ripple, or domino, effects they have on employment throughout the economy. They represent the leading edge of our economy that is built on the ingenuity of the American people and their future growth is increasingly dependent on effective protection of IP rights both here and abroad.

Chapter 2 - Intellectual property rights (IPR) traditionally have been matters of national concern. Individual nation states have developed IPR regimes reflecting their domestic needs and priorities. Over time, intellectual property protection and enforcement have come to the forefront as a key international trade issue for the United States, figuring prominently in the multilateral trade policy arena and in regional and bilateral U.S. free trade agreements (FTAs).

The protection and enforcement of IPR in the United States and abroad is of key interest to Congress. Intellectual property is an increasingly critical component of the U.S. economy. Industries that rely on intellectual property protection in the United States claim to lose billions of dollars each year due to overseas IPR infringement. In light of the recent international financial crisis and global economic downturn, congressional interest has grown in the role of IPR in advancing U.S. industrial competitiveness and contributing to U.S. economic recovery following the recent international crisis. Members of Congress also have expressed concern about the potential health and safety consequences of counterfeit pharmaceutical drugs and other products, as well as the possible link between terrorist groups and traffic in counterfeit and pirated goods.

This chapter discusses the different kinds of IPR and forms of IPR infringement; importance of IPR to the U.S. economy; estimated losses associated with IPR infringement; organizational structure of IPR protection in multilateral, regional, bilateral arenas; U.S. government agencies involved with IPR and trade; and issues for Congress regarding IPR and international trade.

In: Intellectual Property
Editor: Nicolas Vallée

ISBN: 978-1-62417-313-4

Chapter 1

INTELLECTUAL PROPERTY AND THE U.S. ECONOMY: INDUSTRIES IN FOCUS*

Economics and Statistics Administration and U.S. Patent and Trademark Office

FOREWORD

As President Obama has said, if we are to win the future and be successful in an increasingly competitive international market, the United States of America must innovate. Innovation, the process through which new ideas are generated and put into commercial practice, is a key force behind U.S. economic growth and national competitiveness. Likewise, U.S. companies' use of trademarks to distinguish their goods and services from those of competitors enables firms to capture market share and further strengthen our economy. The dynamics of a globally connected market mean that the United States will need to develop the brightest minds with the most advanced training to make the best products.

The Obama Administration's determination to promote innovation and protect intellectual property (IP) rights will harness the inherent drive and ingenuity of the American people in meeting that goal.

---

* This is an edited, reformatted and augmented version of Economics and Statistics Administration and United States Patent and Trademark Office publication, dated March 2012.

Innovation protected by IP rights is key to creating new jobs and growing exports. Innovation has a positive pervasive effect on the entire economy, and its benefits flow both upstream and downstream to every sector of the U.S. economy. Intellectual property is not just the final product of workers and companies—every job in some way, produces, supplies, consumes, or relies on innovation, creativity, and commercial distinctiveness. Protecting our ideas and IP promotes innovative, open, and competitive markets, and helps ensure that the U.S. private sector remains America's innovation engine.

There is a broad range of industries that benefit from IP, both directly and indirectly, and for every innovation in a given industry, generally there are corresponding economic opportunities for other industries to bring advances to the public. Examples of these complementary industries include the computer manufacturer that uses inputs made by semiconductor firms to make the hardware that is needed to run applications made by software companies, the Internet company that generates the on-line applications to distribute copyrighted music, and the auto manufacturer that incorporates patented energy-saving engines in the cars it sells. In each of these cases, industries are supported by the complementary products and services of another industry, so each industry is in a position to benefit from the safeguard of the IP underpinning their business models. When companies are more confident that their ideas will be protected, they have the incentive to pursue advances that push efficiency forward, costs down, and employment up.

Further evidence of this domino effect is seen as downstream businesses benefit from innovative products that lower their costs and improve their processes and finished articles. For example, a more precise machine could make manufacturing pharmaceuticals safer; a more reliable software program could improve military hardware while making it more affordable; or a clearer sound system could boost the quality of a recording studio's album. In each of these cases, and many more, the innovation feeds directly into a new finished article or service that has commercial value. The innovation may increase customer satisfaction through a higher quality product or service or decrease production costs. This additional value can help businesses protect earnings that can support their labor force.

Finally, in addition to the companies that produce, complement, and consume the innovation, logistical and supporting businesses are required to keep these companies up and running. These entities include marketing firms that inform consumers about innovations that can save time and money, packaging companies that ship advanced goods to retailers, and Internet service providers that create and maintain the communications pathways needed to compete in an increasingly on-line world.

This chapter by the Economics and Statistics Administration (ESA) and the U.S. Patent and Trademark Office (USPTO) of the U.S. Department of Commerce attempts to identify the first-order players that are growing IP in the U.S. economy and protecting their innovations through patents, trademarks, or copyrights. These IP-intensive industries support tens of millions of jobs and contribute several trillion dollars to our gross domestic product (GDP). This report not only estimates the contributions of these industries to our economy, but also gauges the ripple, or domino, effects they have on employment throughout the economy. They represent the leading edge of our economy that is built on the ingenuity of the American people and their future growth is increasingly dependent on effective protection of IP rights both here and abroad.

Rebecca M. Blank
Acting Deputy Secretary of Commerce And Under Secretary for
Economic Affairs, Economics and Statistics Administration

David J. Kappos
Under Secretary for Intellectual Property and Director, U.S. Patent and
Trademark Office

EXECUTIVE SUMMARY

Innovation—the process through which new ideas are generated and successfully introduced in the marketplace—is a primary driver of U.S. economic growth and national competitiveness.[1] Likewise, U.S. companies' use of trademarks to distinguish their goods and services from those of competitors represents an additional support for innovation, enabling firms to capture market share, which contributes to growth in our economy. The granting and protection of intellectual property rights is vital to promoting innovation and creativity and is an essential element of our free-enterprise, market-based system. Patents, trademarks, and copyrights are the principal means used to establish ownership of inventions and creative ideas in their various forms, providing a legal foundation to generate tangible benefits from innovation for companies, workers, and consumers. Without this framework, the creators of intellectual property would tend to lose the economic fruits of their own work, thereby undermining the incentives to undertake the investments necessary to develop the IP in the first place.[2] Moreover, without IP protection, the inventor who had invested time and money in developing the new product or service (sunk costs) would always be at a disadvantage to the new firm that could just copy and market the product without having to recoup any sunk costs or pay the higher salaries required by those with the creative talents and skills. As a result, the benefits associated with American ingenuity would tend to more easily flow outside of the United States.

IP is used everywhere in the economy, and IP rights support innovation and creativity in virtually every U.S. industry. While IP rights play a large role in generating economic growth, little attention has been given to identifying which industries produce or use significant amounts of IP and rely most intensively on these rights. This report begins such an investigation by developing several industry-level metrics on IP use and employing these measures to identify a set of the most IP-intensive industries in the U.S. economy. To develop the industry-level metrics discussed, several databases were used, some of which (for the patent and trademark analyses) are publicly available.[3] In the future, more user-friendly sets of these patent and trademark data will be made available on the U.S. Patent and Trademark Office (USPTO) website.

This report employs USPTO administrative data to identify the industries that most intensively use the protection offered by patents and trademarks. For copyrights, the report identifies the set of industries primarily responsible for both the creation and production of copyrighted materials. The report then uses standard statistical methods to identify which American industries are the most patent-, trademark-, and copyright-intensive, and defines this subset of industries as "IP-intensive." Using data collected from sources across the U.S. government, the report examines both the important trends and economic characteristics of these highly IPintensive industries and their meaningful contributions to the U.S. economy. There are several important findings contained in the report.

According to the analysis in this report, the direct and indirect employment in these industries is substantial: Direct employment in the subset of most IP-intensive industries identified in this report amounted to 27.1 million jobs in 2010, while indirect activities associated with these industries provided an additional 12.9 million jobs throughout the economy in 2010, for a total of 40.0 million jobs, or 27.7 percent of all jobs in the economy.

Because all U.S. industries rely on IP to some degree, the statistics reported here for the sectors that use IP most intensively may tend to under-represent the broad impact of IP in the American economy. Moreover, the statistics reported here may not fully reflect the long-run economic benefits and costs of IP in promoting innovation and productivity growth. For example, while this report shows that employment in trademark-intensive industries is almost six times as great as employment in patent-intensive industries, it may be that the kinds of innovation protected by patents play a larger role in driving the long-run growth of productivity throughout the economy.

This report does not contain policy recommendations and is not intended to directly advance particular policy issues. By developing new quantitative measures of IP-intensity by industry, the report aims to promote a better understanding of the industries where IP plays a particularly important role. Although policy issues are not discussed in this report, as a general matter, we note the importance of achieving a balanced system of IP rights that protects inventors and creators from unlawful use of their work while encouraging innovation, competition, and the markets for technology in which IP is transacted. Importantly, using IP rights to support innovation and creativity means recognizing the public domain and limits such as fair use which balance the public's right to use content legally with IP owners' interests.

Principal Findings

- The entire U.S. economy relies on some form of IP, because virtually every industry either produces or uses it.
- By focusing on relevant data and various statistical measures, this report identified 75 industries (from among 313 total) as IP-intensive. These IP-intensive industries directly accounted for 27.1 million American jobs, or 18.8 percent of all employment in the economy, in 2010.[4] A substantial share of IP-intensive employment in the United States was in the 60 trademark-intensive industries, with 22.6 million jobs in 2010. The 26 patent-intensive industries accounted for 3.9 million jobs in 2010, while the 13 copyright-intensive industries provided 5.1 million jobs.[5]
- IP-intensive industries accounted for about $5.06 trillion in value added, or 34.8 percent of U.S. gross domestic product (GDP), in 2010.
- While IP-intensive industries directly supported 27.1 million jobs either on their payrolls or under employment contracts, these sectors also indirectly supported 12.9 million more supply chain jobs throughout the economy. In other words, every two jobs in IP-

intensive industries support an additional one job elsewhere in the economy. In total, 40.0 million jobs, or 27.7 percent of all jobs, were directly or indirectly attributable to the most IPintensive industries.

- Due primarily to historic losses in manufacturing jobs, overall employment in IP-intensive industries has lagged other industries during the last two decades. While employment in non-IP-intensive industries was 21.7 percent higher in 2011 than in 1990, overall IP-intensive industry employment grew 2.3 percent over this same period. Because patent-intensive industries are all in the manufacturing sector, they experienced relatively more employment losses over this period, especially during the past decade. While trademark-intensive industry employment had edged down 2.3 percent by the end of this period, copyright-intensive industries provided a sizeable employment boost, growing by 46.3 percent between 1990 and 2011.
- Between 2010 and 2011, the economic recovery led to a 1.6 percent increase in direct employment in IP-intensive industries, faster than the 1.0 percent growth in non-IP-intensive industries. Growth in copyright-intensive industries (2.4 percent), patent-intensive industries (2.3 percent), and trademark-intensive industries (1.1 percent) all outpaced gains in non-IP-intensive industries.
- Jobs in IP-intensive industries pay well compared to other jobs. Average weekly wages for IP-intensive industries were $1,156 in 2010 or 42 percent higher than the $815 average weekly wages in other (non-IP-intensive) private industries. This wage premium nearly doubled from 22 percent in 1990 to 42 percent by 2010. Patent- and copyright-intensive industries have seen particularly fast wage growth in recent years, with the wage premium in patent-intensive industries increasing from 66 percent in 2005 to 73 percent in 2010, and the premium in copyright-intensive industries rising from 65 percent to 77 percent.
- The comparatively high wages in IP-intensive industries correspond to, on average, the completion of more years of schooling by these workers. More than 42 percent of workers aged 25 and over in these industries in 2010 were college educated, compared with 34 percent on average in non-IP-intensive industries.
- Merchandise exports of IP-intensive industries totaled $775 billion in 2010, accounting for 60.7 percent of total U.S. merchandise exports.

- Data on foreign trade of IP-intensive service-providing industries is limited; however, this report does find that exports of IP-intensive service-providing industries accounted for approximately 19 percent of total U.S. private services exports in 2007.

I. INTRODUCTION

"The key to our success – as it has always been – will be to compete by developing new products, by generating new industries, by maintaining our role as the world's engine of scientific discovery and technological innovation.
It's absolutely essential to our future."

– President Barack Obama, November 17, 2010.

A defining characteristic of the U.S. economy throughout its history is its rapid and sustained growth. For example, real income per person two hundred years ago was four percent of what it is today.[6] Strong and sustained economic growth results from several factors, but among the most important is innovation, "the process by which individuals and organizations generate new ideas and put them into practice" and create "new and better ways of producing goods and services."[7] Likewise, U.S. companies' use of trademarks to distinguish their goods and services from those of competitors represents an additional support for innovation, enabling firms to capture market share and promoting growth in our economy.

One important way to help encourage innovation is through the protection of intellectual property (IP). The investments necessary to develop IP are often quite substantial. Firms and individuals, in order to invest the necessary resources, need some assurance that they will benefit from and recover the costs of the creation of intellectual property. IP rights help protect authors, inventors, and merchants of goods and services from having their creations and innovations quickly and easily exploited by other firms or individuals, diminishing the benefits to the inventor of the IP. This reduction in private benefits to be gained from the underlying innovation could, in turn, reduce the incentives to undertake the investments necessary to develop the IP in the first place.[8]

Protection of IP has been a critical function of the U.S. government since the founding of this country. Indeed, Article I, Section 8, Clause 8 of the U.S.

Constitution grants to Congress the power to "promote the Progress of Science and useful Arts by securing for limited Times to Authors and Inventors the exclusive Right to their respective Writings and Discoveries."

Patents, trademarks, and copyrights are the principal means for establishing ownership rights to inventions and ideas, and they provide a legal foundation by which intangible ideas generate tangible benefits to firms and workers. IP protection affects commerce throughout the economy by:

- Providing incentives to invent and create;
- Protecting innovators from unauthorized copying;
- Facilitating vertical specialization in technology markets;
- Creating a platform for financial investments in innovation;
- Supporting entrepreneurial liquidity through mergers, acquisitions, and IPOs;
- Making licensing-based technology business models possible; and
- Enabling a more efficient market for technology transfer and trading in technology and ideas.

Certain industries find IP protection to be especially important and consequently register a relatively high number of patents, trademarks, and copyrights compared with other similarly sized industries. Still, as a general matter, we note the importance of achieving a balanced system of IP rights that protects inventors and creators from unlawful use of their work while encouraging innovation, competition, and the markets for technology in which IP is transacted. Importantly, using IP to support innovation and creativity means recognizing the public domain and limits such as fair use which balance the public's right to use content legally with IP owners' interests.

This chapter, prepared by the Economics and Statistics Administration (ESA) and the U.S. Patent and Trademark Office (USPTO) of the U.S. Department of Commerce, focuses on identifying these IP-intensive industries and examining their characteristics and contributions to the overall economy. Using administrative data and economic literature analyzing IP protection, this report develops various metrics of IP usage by industry. The report does not contain policy recommendations and is not intended to directly advance particular policy issues.

By developing new measures of IP-intensity by industry, the report provides a better understanding of the industries where IP plays a particularly important role.

To be specific, the report identifies 75 industries (from among 313 total) that are particularly dependent on patent, copyright, or trademark protection and which we define as IP-intensive. The second section of this report, "Identifying IP-Intensive Industries," lays out the methodologies used to develop these metrics and details which industries were selected as IP-intensive.

Given the differing natures of patents, copyrights and trademarks along with data limitations, we developed separate approaches that are consistent, to the extent possible, for identifying which industries rely on patents, trademarks, or copyrights.

In the third section of this report, we present evidence on the size of these industries relative to the economy as a whole. This analysis points to several key results. First, these industries accounted for 27.1 million, or 18.8 percent, of all jobs in 2010.[9]

They contributed 34.8 percent to gross domestic product (GDP), with total value added of $5.06 trillion in 2010. The bulk of employment and value added correspond to the 60 trademark-intensive industries, which is a reflection of the nearly ubiquitous use of trademarks and logos in the marketplace. The share of total employment in many IP-intensive industries has edged down over the past two decades. Further, it is important to note that these 75 industries are by no means the only ones with business models and profits supported by IP protection. Indeed, as discussed more below, this report reinforces the notion that all sectors of the economy actively use IP protection.

In addition, we examine how these industries have performed relative to other, less IP-intensive industries. In particular, although IP-intensive industries have not performed as well in job growth relative to other sectors over the last two decades, IP-intensive industries have been a source of high-quality jobs, with average weekly wages in these industries averaging 42 percent higher than the average for all other industries. The workers in IP-intensive industries also tend to be better educated, on average, than other workers: more than 42 percent of workers age 25 and over in these IP-intensive industries were college educated, compared to one in three in other industries.

Copyright-intensive industries, in particular, employed the most educated workers, with 61 percent holding at least a college degree. Finally, IP-intensive industries awere large contributors to U.S. foreign trade, accounting for about 60.7 percent of total merchandise exports and about 19 percent of service exports.

II. Identifying IP-Intensive Industries

Patents

Overview

Utility patents, hereafter referred to as "patents" unless otherwise noted, assist owners in protecting the rights to inventions and innovative processes.[10] A patent issued by the USPTO enables the owner to pursue legal action to exclude "others from making, using, offering for sale, or selling the invention throughout the United States or importing the invention into the United States."[11] U.S. patents are issued to inventors who can assign their ownership rights to individuals, corporations, universities, other organizations, or branches of governments of any nationality.[12, 13] Our patent analysis focuses only on patents issued to U.S corporations, which accounted for about 45 percent of total patents issued between fiscal years (FY) 2004 and 2008 and 87 percent of all U.S.-owned patents for this time period.[14, 15]

Patents are further classified in over 430 patent "technology classes" that distinguish their inventive content.[16] Additionally, the USPTO maintains a general concordance between its technology classifications and 32 North America Industry Classification System (NAICS) codes (26 unique codes and 6 combinations), which enables analysts to associate patents with these industries.[17] We rely on these NAICS-based patent counts for FY 2004 to FY 2008 to identify patent-intensive industries.18 Just as a patent can be assigned to more than one technology class, it also may be associated with multiple industries. Because no similar concordances to NAICS are available for plant or design patents, only utility patents were used in our analysis.19 This approach strictly limits the patent analysis to the manufacturing sector because the concordance system only associates patents with manufacturing industries. Service-providing industries may also rely on utility patents in their production processes, but these industries are not captured by the patent-NAICS concordance that we employ.

Fractional vs. Whole Patents Counts

The USPTO reports patent data by NAICS category using two different counting measures. The first gives one full count to every industry associated with a particular patent. The second divides each patent by the number of industries it is associated with, thus creating fractional counts of patents. The sum of the fractional counts equals the total number of patents issued in a given year, while the sum of the whole counts across industries is greater than the number of patents issued. Patent analyses within a given industry or technology class commonly use whole counts; however, cross-industry analyses typically use fractional counts in order to avoid over-counting. For these same reasons, fractional counts are used in this report.[20, 21] It is important to note that the NAICS concordance maintained by the USPTO associates each patent with its final use in the economy. In practice, this means that the patents are all associated with manufacturing NAICS codes regardless of whether the company that owns them is a manufacturer or a company that may be classified in the services sector.[22]

Methodology

The USPTO has NAICS-based patent data covering the period from 1963 to 2008. We calculated a measure of industry patent "intensity," defined as the ratio of total patents over the five years in a NAICS category to the average payroll employment by industry. Because employment is a gauge of industry size, dividing patent counts by employment normalizes patenting activity with respect to industry size.[23] This approach helps put all industries on an even playing field, so that the most patent-intensive industries were defined not as the ones with the most patents, but rather those with the most patents per job.

By using a five-year period (in this case, FY 2004-08, the most recent period for which data are available) instead of just one year helps minimize the chance that anomalies in any given year will skew our results. The analysis was performed at the greatest possible level of NAICS industry detail, and so results include four-digit industries as well as some individual three-digit industries and combinations of three- or four-digit industries.[24] As will be seen in the results, most patent-intensive industries in the sample fall into the four-digit NAICS industries, which may be a product of the patent-intensive nature of these more finely disaggregated industry sectors.

Results

We defined patent-intensive industries as ones with above-average patent intensity (patent/job ratio) when comparing all industries.[25] (See Table 1.) The four most patent-intensive industries all have intensity rates that are one standard deviation above the mean patent-intensity cutoff, and are all classified in computer and electronic product manufacturing (NAICS 334). This three-digit NAICS industry includes computer and peripheral equipment; communications equipment; other computer and electronic products; semiconductor and other electronic components; and navigational, measuring, electro-medical, and control instruments.

This is unsurprising when one also looks at the recent top ten U.S. companies ranked by granted patents.[26, 27] This group of companies includes Intel, Hewlett-Packard, Micron Technology, and Texas Instruments, each of which is closely associated with computer and computer peripheral manufacturing.

USPTO's Patent Class - NAICS Concordance

The scheme used by the USPTO to associate utility patents with their proper NAICS designation includes groups of four-digit as well as three-digit NAICS codes.

With respect to machinery manufacturing (NAICS 333) and electrical equipment, appliance, and component manufacturing (NAICS 335), the USPTO does not provide a breakdown beyond this three-digit level, largely because many of the classes in these categories overlap with other NAICS codes.

Thus, our analysis of patents cannot measure patent intensity in NAICS 333 and 335 at a more disaggregated level because technology classes corresponding to these industry categories cannot easily and cleanly be associated with more detailed NAICS industries. Because we cannot disaggregate these two industries, or the groupings of NAICS 3343 and 3346, and NAICS 3253, 3255, and 3256 (which also were found to be patent-intensive) into their four-digit NAICS components, it is not possible to know whether each of the individual four-digit industries is patent-intensive according to our definition.

Alternatives and Robustness

The methodology used in this report identifies industries only, and so should not be used to draw conclusions about specific companies in these industries, or in other industries which did not make our list.

Table 1. Patent Intensity, FY 2004-08

Above mean

NAICS code	Industry title	Patents (number)	Employment (1000 jobs)	Patent intensity (patents/ 1000 jobs)
3341	Computer and peripheral equipment	54,416	196.1	277.5
3342	Communications equipment	35,797	135.2	264.8
3344	Semiconductor and other electronic components	50,088	448.7	111.6
3343,-6	Other computer and electronic products	7,744	71.4	108.5
3345	Navigational, measuring, electromedical, and control instruments	42,415	441.3	96.1
3251	Basic chemicals	12,109	150.9	80.2
335	Electrical equipment, appliance, and components	23,503	433.0	54.3
3254	Pharmaceutical and medicines	13,627	291.3	46.8
3399	Other miscellaneous	12,717	339.2	37.5
3253,-5,-6,-9	Other chemical products and preparation	10,322	318.1	32.4
3391	Medical equipment and supplies	9,716	303.2	32.0
333	Machinery	37,105	1,173.7	31.6
3252	Resin, synthetic rubber, fibers, and filaments	2,771	106.4	26.0
326	Plastics and rubber products	8,289	775.8	10.7
3361-3363	Motor vehicles, trailers and parts	8,298	1,029.8	8.1
327	Nonmetallic mineral products	3,651	497.2	7.3
3365,-6,-9	Other transportation equipment	1,585	222.4	7.1
3364	Aerospace products and parts	2,726	473.3	5.8
313,-4,-5,-6	Textiles, apparel and leather	2,566	632.2	4.1
332	Fabricated metal products	5,495	1,532.5	3.6
331	Primary metals	998	458.9	2.2
337	Furniture and related products	1,107	543.0	2.0
322,-3	Paper, printing and support activities	1,313	1102.6	1.2
312	Beverage and tobacco products	217	195.5	1.1
321	Wood products	428	527.8	0.8
311	Food	719	1,483.1	0.5
Total		354,360	13,882.7	25.5

Source: ESA calculations using data from USPTO and the Bureau of Labor Statistics' Current Employment Statistics program. Note: The patent count is the total for fiscal years 2004-08, while employment shows the calendar year 2004-08 average in thousands of jobs.

For example, industries with a relatively small stock of highly valuable patents, but relatively large numbers of employees, would not be selected as patent-intensive in our methodology, even though companies in that industry may rely critically upon patent protection.

The aerospace industry is a good illustration of this point. Its patent-to-jobs ratio was about one-fifth the overall average, but there is other evidence suggesting that patenting is considered important by aerospace companies. In a survey conducted by researchers at Carnegie-Mellon University in 1994, aerospace company R&D managers reported on average that 78 percent of their business units filed for patents (about the same share reported by the pharmaceutical and basic chemical industries, both of which make our list of patent-intensive industries) and that 51 percent of product innovations were being patented (about the same share as in the semiconductor industry, which was also included in our list).[28]

So, while it does appear that patenting was relatively important to companies in the aerospace industry during the early 1990s, this sector does not show up in our current list.

This outcome could indicate that patent protection is less useful in aerospace than two decades ago (probably unlikely), or it may be an artifact of the manner in which technologies are assigned to industries in our methodology.

Another possibility is that industries may rely on proprietary protections other than patents, such as a trade secret, which may very well be a safer mode of protecting new discoveries in an industry with few competitors monitoring each other closely.

Whereas patenting requires disclosure of the details of the new discovery, a trade secret allows the discovery to remain undisclosed, under close in-house protection.

The Carnegie-Mellon University survey allows the consideration of an alternative measure of IP reliance, based not on the number of patents granted per employee, but instead on the responses of research and development (R&D) lab managers to questions concerning the effectiveness of different methods for protecting innovation. Because managers were asked in the survey to report on the percentage of product and process innovations for which patent protection was effective at capturing competitive advantage from the innovation, the study provides a window into the relative importance of patent protection considered against the stock of innovations that industries are producing.

As Table 2 makes clear, this survey method results in a different ranking among industries than the patent intensity measures used in this report, with sectors such as pharmaceuticals (drugs), medical equipment, and chemicals ranked most highly.

Table 2. Percent of Product and Process Innovations for which Patents were Considered an Effective Mechanism for Appropriating the Returns to Innovation, by Industry

SIC code	Industry title	Product innovations (percent)	SIC code	Industry title	Process innovations (percent)
3311	**Medical equipment**	**54.70%**	**2320**	**Petroleum**	**36.67%**
2423	**Drugs**	**50.20%**	**2423**	**Drugs**	**36.15%**
2920	**Special purpose machinery, nec.**	**48.83%**	**3311**	**Medical equipment**	**34.02%**
3430	**Auto parts**	**44.35%**	**2700**	**Metal, nec.**	**31.67%**
3010	**Computers**	**41.00%**	**2610**	**Glass**	**30.83%**
2429	**Miscellaneous chemicals**	**39.66%**	**3010**	**Computers**	**30.25%**
2800	**Metal products**	**39.43%**	**2411**	**Basic chemicals**	**29.71%**
3410	**Car/truck**	**38.89%**	**2920**	**Special purpose machinery, nec.**	**28.57%**
2411	**Basic chemicals**	**38.86%**	**2100**	**Paper**	**27.58%**
2910	**General purpose machinery, nec.**	**38.78%**	**2429**	**Miscellaneous chemicals**	**27.32%**
3230	**TV/radio**	**38.75%**	**1700**	**Textiles**	**25.22%**
2400	**Chemicals, nec.**	**37.46%**	**3430**	**Auto parts**	**24.35%**
2100	**Paper**	**36.94%**	**2910**	**General purpose machinery, nec.**	**23.62%**
2922	**Machine tools**	**36.00%**	**3600**	**Other manufacturing**	**23.42%**
3100	Electrical equipment	34.55%	**2600**	**Mineral products**	**23.33%**
3600	Other manufacturing	33.81%	**3210**	**Semiconductors and related equipment**	**23.33%**
2320	Petroleum	33.33%	2800	Metal products	22.50%
2413	Plastic resins	32.96%	3110	Motor/generator	22.14%
3530	Aerospace	32.92%	3410	Car/truck	21.67%
2500	Rubber/plastic	32.71%	3530	Aerospace	21.38%
2610	Glass	30.83%	2413	Plastic resins	21.30%
2695	Concrete, cement, lime	30.00%	2400	Chemicals, nec.	20.40%
3314	Search/navigational equipment	28.68%	2500	Rubber/plastic	19.86%
3210	Electronic components	26.67%	3100	Electrical equipment	19.09%
3312	Precision instruments	25.86%	3230	TV/radio	18.75%
3220	Communications equipment	25.74%	2695	Concrete, cement, lime	18.50%
3110	Motor/generator	25.23%	2922	Machine tools	18.00%
2710	Steel	22.00%	3312	Precision instruments	16.77%
3210	Semiconductors and related equipment	21.35%	1500	Food	16.40%
2600	Mineral products	21.11%	2710	Steel	15.50%
1700	Textiles	20.00%	3210	Electronic components	15.19%
2700	Metal, nec.	20.00%	3220	Communications equipment	14.70%
1500	Food	18.26%	3314	Search/navigational equipment	13.24%
2200	Printing/publishing	12.08%	2200	Printing/publishing	8.64%
	All industries	**34.83%**		**All industries**	**23.30%**

Source: Cohen et al. 2000, Tables 1 and 2.

Note: Estimates are the mean percentage of product or process innovations for which patents were considered by respondent to be effective for securing competitive advantage. Bolded industries in each list are those with response scores above the industry-level mean for all industries surveyed in the study. Industries are defined according to the 1987 Standard Industrial Classification (SIC) system, the precursor to NAICS. The abbreviation "nec." stands for "not elsewhere classified."

Because this survey was conducted in the mid 1990s, it was not considered an ideal principal measure for current industry activities.

It is noteworthy, however, that the industries reporting that patents were effective for more than 51.0 percent of product innovations and 23.3 percent of process innovations (the overall averages in the survey) are largely the same industries with above-average patent-to-jobs ratios (patent intensity) in the mid-to-late 2000s.

For this reason, the Carnegie-Mellon University survey results serve as an important robustness check for the accuracy of the methodology selected in this report for patent-intensive industries.

TRADEMARKS

Overview

Industries throughout the economy rely on trademarks registered at the USPTO to protect brands for the goods and services they market.[29] A trademark is “a word, phrase, symbol, or design, or a combination thereof, that identifies and distinguishes the source of the goods of one party from those of others.”[30]

Unlike a patent, which protects an invention, or a copyright, which protects a work of original authorship, a trademark does not protect a new product or service *per se*. A trademark instead confers protection upon the brand or identity of a good, thus preventing competitors from leveraging another firm’s reputation and confusing consumers as to the source of the goods.

Service marks are similar in nature to trademarks, but distinguish the source of a service rather than a good.[31] In the remainder of this text, the term “trademark” will refer to both trademarks and service marks. Any company or individual, American or foreign, can apply to register a trademark with the USPTO for a nominal fee.

Once granted, trademark registrations can remain in force indefinitely as long as the trademark remains in active use and maintenance payments are made. If the trademark is not in active use and maintained, the rights will cease after a period of time, normally five years. To maintain trademark registrations, new owners must file a Section 8 Affidavit of Continuous Use before the end of the sixth year after the initial registration date. Trademark registrations must be renewed on or about every 10-year anniversary of the registration of the trademark.[32]

Identifying Trademark-Intensive Industries

Unlike patents, there is little academic research examining industry use of trademarks. Accordingly, this report offers what may be the first comprehensive analysis of trademark use by U.S. industries that is grounded in original research, data, and measurement theory. As a preliminary matter, we recognized that each trademark registration has a description of the type of good or service with which the protected mark is used in commerce. The USPTO classifies goods and services for administrative convenience, and applicants for trademark registration must provide a separate description—and pay separate application and maintenance fees—for each "class" in which the goods or services associated with the trademark is classified.[33] This makes working with trademark registration data different from working with patent grant data.

Trademark application and maintenance fees are assessed on a per-class basis, and registration holders may elect to renew their registrations with respect to some but not all classes. As a result, holding a "multi-class" registration is practically equivalent to holding multiple registrations, one for each class. Accordingly, in the foregoing analysis each class listed on a registration was considered as the unit of analysis, creating a class-registration count. For example, if one mark (or logo) is registered in three classes, then our input measure in the analysis reflects three trademark registration counts, one for each class.[34]

In practice, the vast majority of trademark registrations are associated with a single class. However, from 2000 to 2009, about 16 percent of trademarks were registered in more than one class. About 10 percent were registered in two classes, while the remaining six percent were registered in anywhere from three to 45 classes. As a result of this variation, the sum of registration-class observations is somewhat greater than the sum of registrations.[35]

As an example of a firm that has chosen to register a trademark in multiple classes, consider Whole Foods' "365 Everyday Value" brand. Whole Foods protects the 365 Everyday Value brand in 10 different trademark classes.[36] The goods classes under which the 365 Everyday Value logo is applied include processed foods (from frozen dinners to trail mix), staple foods (cereal, tortilla chips, pasta sauce, cookies), as well as food storage containers, paper products (paper towels and coffee filters), and cosmetics and cleaning preparations (laundry detergent, shampoo, bath gels). In effect, these 10 classes are 10 distinct registrations, each of which can be renewed, or not, separately.

The approach that we adopted for measuring trademark-intensive industries parallels, but differs from, the approach employed when analyzing patents.[37] For patents, each patent was counted only once overall; for trademark measurements, each mark is counted once for each class in which it belongs, potentially counting it more than once overall according to the number of classes in which it is registered.

Since it is not easy to ascertain which trademark class-registrations are relatively more important, we used the best measure available, based on the economic realities of the fee system.[38] Because each trademark class registration involves some fixed filing fee paid to the USPTO, the more classes in which a trademark is registered indicate more times that a fee has been paid to the USPTO. Using these fee-payments is an effective method to base an IP-intensity measure and this approach was followed consistently for both trademarks and patents. Also, as a sensitivity measure, we found that counting patents and trademarks differently, by using whole counts for patents for example, produced similar results.

Because trademark registrations can be segmented by class but not by industry, there is no USPTO NAICS concordance for these data. Due to this methodological limitation, there is no single, straightforward way to tabulate registrations and measure trademark intensities by industry. Accordingly, and because the measurement of trademark use is a new science, we opted for over-inclusiveness and developed a three-pronged approach to identifying trademark-intensive industries.

We relied on three related but distinct approaches, using different samples of companies that have registered USPTO trademarks. The first approach is the closest approximation to the methodology used to identify patent-intensive industries. Starting with the complete set of trademark registrations, we matched publicly traded companies by their name to a separate database containing information on the firms' primary industry and number of employees.

These data allowed us to calculate trademark intensities by industry for the matched firms. In the second approach, we reviewed lists of the top 50 corporate trademark registrants published by the USPTO (which, unlike the first approach, include both private and public companies) and identified industries that appear repeatedly. To help moderate the tendency of the first two approaches to under-represent smaller and younger firms, our third approach focused on a representative and random sample of firms drawn from the complete database of U.S. corporate trademark registrations from FY 2004 to FY 2008.

Trademark Intensities

Methodology

Parallel to our method for defining patent intensities, we defined trademark intensities as the ratio of trademark registrations to employment in a given industry. Thus, we measured the number of trademark registrations per employee. The USPTO applied a firm-name standardization routine developed originally for patent analysis in order to match companies with new trademark registrations to companies in Compustat's database of financial statements of publicly traded companies.[39] This matching identified 386,998 distinct standardized firm names in the FY 2004-08 trademark registration and trademark renewal records. These were firms that either registered or renewed at least one trademark from FY 2004 to FY 2008. These records were matched with 9,539 parent-company records drawn from the Compustat database.

Successful matches were made for 3,475 firms. Since Compustat records also include NAICS codes, it was straightforward to sum trademark registrations and employment by four-digit NAICS industry and then estimate industry trademark intensity as the number of matched firm trademarks per worker in each industry.[40]

Because Compustat does not record the relationships between parent and subsidiary companies, the identities of trademark registrants were only matched to the name of the publicly traded parent company reporting financial statements to the U.S. government. Accordingly, trademarks registered in the name of subsidiaries that have different names than the parent company were not matched by this method.[41]

Results

Overall, this procedure identified 235 four-digit NAICS industries with trademark registrations among publicly traded parent companies. The employment-weighted mean trademark intensity for these firms was 1.86 trademarks per 1,000 workers, and the weighted standard deviation was 3.80. Upon examining this distribution, we defined as trademark-intensive the 55 industries with trademark intensities above the sample industry mean, consistent with our approach for identifying patent-intensive industries. We also set as a minimum a sample size in each four-digit NAICS industry of five distinct firms and 100 registrations over the 5-year period. These minima helped us avoid wrongly selecting as trademark-intensive any industries that had high estimated trademark intensities but very few firms in the matched sample or very low absolute trademark registration activity.

As shown in Table 3, the 55 industries span most major industry sectors, with many industries in the manufacturing and information sectors, as well as others from the financial activities, professional and technical services, mining, construction, healthcare, and leisure and hospitality sectors. Four industries stood out from the rest due to their notably high trademark intensities: audio and video equipment manufacturing (82.5 trademarks per 1,000 employees), other miscellaneous manufacturing (64.5), satellite telecommunications (35.3), and lessors of nonfinancial intangible assets (33.3). The first two industries also ranked above the mean in our patent-intensive listing. The category "lessors of nonfinancial intangible assets" is a special case. That industry's sole function is to assign "rights to assets, such as patents, trademarks, brand names, and/or franchise agreements [but not to copyrighted works] for which a royalty payment or licensing fee is paid to the asset holder."[42] Thus, the industry is by definition IP- intensive, but could not be selected as patent-intensive because all patents were assigned by the USPTO NAICS concordance only to manufacturing industries.

Top 50 Trademark-Registering Companies

Methodology

Since 2006, the USPTO's annual *Fiscal Year Performance and Accountability Reports* have identified the 50 companies that obtained the largest number of trademark registrations during the Federal fiscal year.[43] Because these reports do not provide the primary NAICS code of the companies in question, we used OneSource, a subscription-based dataset, to determine the relevant NAICS industry classification for each of the companies listed for FY 2006-10.[44, 45] On the theory that these large companies operate in industries with large numbers of trademark registrations relative to the size of the industry, a subset of those industries was defined as trademark-intensive. After assigning each company to a four-digit NAICS industry, we tabulated the number of times each industry appeared in the top 50 during the FY 2006-10 period. We designated industries as trademark-intensive if they had counts of five or higher. At the extremes, counts of five or higher occurred if a given company was among the top 50 registrants in each of the five years studied, or if five different companies in an industry had reached the top 50 in a single year during the period. The tabulations showed neither of those extremes. Two industries were selected based on their score of five—sugar and confectionary products (NAICS 3113) and household appliances (3352)—and in both cases, two or more companies accounted for the top 50 trademark registrations.

Table 3. Industries with Above-Average Trademark Intensity, FY 2004-08

NAICS code	Industry title	Trademark intensity (trademarks/ 1000 workers)	NAICS code	Industry title	Trademark intensity (trademarks/ 1000 workers)
3343	Audio & video equipment mfg.	82.5	3231	Printing & related support activities	3.6
3399	Other miscellaneous manufacturing	64.5	5416	Management & technical consulting	3.5
5174	Satellite telecommunications	35.3	3345	Electronic instrument mfg.	3.4
5331	Lessors of nonfinancial intangible assets	33.3	5241	Insurance carriers	3.4
5191	Other information services	14.8	4234	Commercial equip. merchant wholesalers	3.1
5615	Travel arrangement & reservation	13.5	3115	Dairy product manufacturing	3.0
5179	Other telecommunications	12.4	3222	Converted paper product mfg.	2.9
5311	Lessors of real estate	11.2	3251	Basic chemical manufacturing	2.9
5112	Software publishers	8.2	4511	Sporting goods, hobby, and musical instrument stores	2.9
4541	Electronic shopping & mail-order houses	7.7	3121	Beverage manufacturing	2.8
3256	Soap, cleaning compound, & toiletries	7.4	5614	Business support services	2.7
3322	Cutlery & handtool manufacturing	7.3	3221	Pulp, paper, & paperboard mills	2.6
3339	Other general purpose machinery manufacturing	6.1	3252	Resin, rubber, & artificial fibers	2.6
3391	Medical equipment & supplies manufacturing	5.9	3261	Plastics product manufacturing	2.6
5111	Newspaper, book, & directory publishers	5.8	3259	Other chemical product & preparations	2.5
3333	Commercial & service industry manufacturing	5.4	4242	Druggists' goods merchant wholesalers	2.5
5417	Scientific research & development	5.4	3353	Electrical equipment mfg.	2.4
3359	Other electrical equipment & components	4.7	6214	Outpatient care centers	2.4
3254	Pharmaceutical & medicine mfg.	4.6	5239	Other financial investment activities	2.3
3162	Footwear manufacturing	4.4	7132	Gambling industries	2.3
5121	Motion picture & video industries	4.3	3112	Grain & oilseed milling	2.2
3371	Household & institutional furniture	4.2	3336	Turbine & power transmission equipment	2.2
4244	Grocery & related product wholesalers	4.1	5152	Cable & other subscription programming	2.2
2361	Residential building construction	4.0	3119	Other food manufacturing	2.0
3352	Household appliance manufacturing	3.8	3329	Other fabricated metal product manufacturing	1.9
2111	Oil & gas extraction	3.7	3341	Computer & peripheral equipment	1.9
3332	Industrial machinery manufacturing	3.7	5221	Depository credit intermediation	1.9
3342	Communications equipment mfg.	3.7			

Source: USPTO calculations using the agency's trademark registration data and company employment data from Compustat.

Results

There are 14 industries with counts of five or greater, with other miscellaneous manufacturing at the top with 32.[46] (See Table 4.) Also scoring in the double-digits are pharmaceutical and medicine manufacturing; soap, cleaning compound, and toiletry product manufacturing; and motion picture and video industries.[47] The industries that were selected using this approach further validated the trademark-intensity approach, as 10 of the 14 industries identified here were also flagged based on their trademark intensity. The four additional industries are sugar and confectionary product manufacturing, motor vehicle manufacturing, grocery stores, and radio and television broadcasting.

Table 4. Industries with Five or More Appearances in the Listings of Top 50 Trademark Registering Companies, FY 2006-10

NAICS code	Industry title	Number of Top 50 appearances
3399	Other miscellaneous manufacturing	32
3254	Pharmaceutical and medicine manufacturing	30
3256	Soap, cleaning compound, and toiletry product manufacturing	29
5121	Motion picture and video industries	14
5111	Newspaper, periodical, book, and directory publishers	9
5151	Radio and television broadcasting	9
5241	Insurance carriers	7
3343	Audio and video equipment manufacturing	7
3121	Beverage manufacturing	7
7132	Gambling industries	6
4451	Grocery stores	6
3361	Motor vehicle manufacturing	6
3352	Household appliance manufacturing	5
3113	Sugar and confectionery product manufacturing	5

Source: ESA calculations using data from the USPTO Annual Reports, Table 29B and OneSource.

Random Sample of Trademark Registrations

Methodology

One shortcoming of identifying industries based on the trademark intensities or top 50 appearances is that these approaches tend to bias selection toward larger companies that register a greater number of trademarks. Moreover, these approaches can fail to account for the critical importance that single trademarks may hold for large entities (for instance, Coca-Cola soft drinks) or differences in industry composition and concentration. These methods also can miss industries composed of smaller companies that may account for many trademarks as a group but do not otherwise fall in the USPTO top 50 listing. To help overcome these shortcomings, we supplemented the analyses with a random sample of registrants drawn from the universe of all 166,844 trademarks registered in FY 2010. These data were generated by USPTO using publicly available source data.[48] They comprise all 106,560 trademark registrants, both corporate and individual, including company/person name and number of trademarks registered in that year. To measure the industry share of total registered trademarks, a random sample of 300 registrations was drawn from this dataset. U.S. companies were listed as the registrant on 196 of these 300 trademark registrations, or about two-thirds.[49] We assigned four-digit NAICS industry codes to these firms using the same procedure employed for the top 50 corporate registrants.[50] Although limited in size, the sample's industry distribution reinforces the breadth and depth of trademarking activity seen in the other data highlighted throughout this report.

One limitation of this approach is that the sample was drawn from records that pertain to only a single year. This may result in the under-identification of industries that generally register frequently in an average year, but, for some reason, were less active in 2010. The sample size is another important limitation. The sample we used was small because assigning NAICS codes to each company requires the use of many resources. In an ideal data world, the USPTO trademark registry would include each corporate registrant's NAICS code; however, such information is not provided in trademark applications and therefore is not included in the USPTO database. The process of assigning NAICS codes to companies was especially cumbersome because many of the firms drawn from the sample were small businesses with few employees and little publicly available information from which to find or infer a NAICS classification.

Results

Table 5 below lists the NAICS industries as well as the share of registrations within the sample for all industries with at least five registrations that were drawn from the sample. We recognize that setting the limit at five registrations is somewhat arbitrary. Because there were no natural breaks in the sample distribution of trademark registrations, we chose this cutoff because it was two standard deviations above the mean and thus a relatively high bar.

The seven industries listed in Table 5 account for one-fifth of trademark registrations from the 196 U.S. companies in our sample. Consistent with the list of trademark-intensive industries selected thus far, the ones selected based on the sample span a number of major industry sectors, including manufacturing, retail trade, wholesale trade, and professional and business services.

Furthermore, six of the seven selected industries had already been selected as trademark-intensive based on their trademark intensities. The sole addition is clothing stores, which was one of the two NAICS categories with the highest share of registrations in the sample; the other was management, scientific, and technical consulting services. The selection of these seven industries will be further reinforced through a review of corporate brand rankings and trademark class data.

Table 5. Percent Distribution of Trademark Registrations of Selected Industries from Sample of U.S.-Owned, Trademark-Registrant Companies, Ranked by Percent, FY 2010

NAICS code	Industry title	Percent
4481	Clothing stores	3.5%
5416	Management, scientific, and technical consulting services	3.5%
3121	Beverage manufacturing	2.5%
4234	Professional and commercial equipment and supplies merchant wholesalers	2.5%
4511	Sporting goods, hobby, and musical instrument stores	2.5%
5112	Software publishers	2.5%
5417	Scientific research and development services	2.5%

Source: ESA calculations using trademark registration data provided by USPTO and OneSource.

Identifying Trademark-Intensive Industries

This combined approach for determining trademark intensity by industry—and any comprehensive examination of the USPTO trademark registration database—shows that a wide range of industries use trademarks in commerce.

This fact maps onto our everyday experience, in which trademarks, brands, and logos are ubiquitous in the marketplace.

The results also show that the latter two approaches tend to complement the first approach. The first approach is the most data intensive, and relies on a systematic matching of the entire U.S. trademark registration dataset over a five-year period to a comprehensive dataset of all publicly traded companies in the U.S. Nevertheless, it has limitations, as it may undercount some industries and smaller companies, either because a single trademark in an industry that is highly concentrated may be underrepresented (for instance, Coca-Cola and Pepsi-Cola) or because many small companies are not publicly traded. While the second and third approaches were meant to mitigate these shortcomings, it is interesting to note that there was substantial correlation in the results, with the majority of industries chosen in these two methods also being selected using the criteria in the first analysis.

Nevertheless, our intention to capture a broad swath of trademark-intensive industries and the recognition that there were conceptual shortcomings in each approach has led us to treat all three approaches as complementary, and to define as trademark-intensive any industry identified by any of the three approaches. As will be detailed, the final results are broadly corroborated by a separate review of trademark class data and corporate brand rankings.

Table 6 shows the complete list of 60 trademark-intensive industries, which again are defined as any four-digit NAICS industry that is selected based on any one of the three methods: trademark intensity, top 50 registrants, or trademark registration sample.

All but five of the industries listed in Table 6 were selected based on their above-average trademark intensity. The five additional industries were selected based on either their top 50 registrations or the trademark registration sample. Only one industry, beverage manufacturing (NAICS 3121), was selected through all three methods.

Table 6. Trademark-Intensive Industries and Selection Criteria

NAICS code	Industry title	Selection criteria		
		Trademark intensity	Top 50	Sample
2111	Oil and gas extraction	X		
2361	Residential building construction	X		
3112	Grain and oilseed milling	X		
3113	Sugar and confectionery product manufacturing		X	
3115	Dairy product manufacturing	X		
3119	Other food manufacturing	X		
3121	Beverage manufacturing	X	X	X
3162	Footwear manufacturing	X		
3221	Pulp, paper, and paperboard mills	X		
3222	Converted paper product manufacturing	X		
3231	Printing and related support activities	X		
3251	Basic chemical manufacturing	X		
3252	Resin, rubber, and artificial fibers	X		
3254	Pharmaceutical and medicine manufacturing	X	X	
3256	Soap, cleaning compound, and toiletries	X	X	
3259	Other chemical product and preparations	X		
3261	Plastics product manufacturing	X		
3322	Cutlery and handtool manufacturing	X		
3329	Other fabricated metal product manufacturing	X		
3332	Industrial machinery manufacturing	X		
3333	Commercial and service industry manufacturing	X		
3336	Turbine and power transmission equipment	X		
3339	Other general purpose machinery manufacturing	X		
3341	Computer and peripheral equipment	X		
3342	Communications equipment manufacturing	X		
3343	Audio and video equipment manufacturing	X	X	
3345	Electronic instrument manufacturing	X		
3352	Household appliance manufacturing	X	X	
3353	Electrical equipment manufacturing	X		
3359	Other electrical equipment and components	X		
3361	Motor vehicle manufacturing		X	
3371	Household and institutional furniture	X		
3391	Medical equipment and supplies manufacturing	X		
3399	Other miscellaneous manufacturing	X	X	
4234	Commercial equip. merchant wholesalers	X		X
4242	Druggists' goods merchant wholesalers	X		

NAICS code	Industry title	Selection criteria		
		Trademark intensity	Top 50	Sample
4244	Grocery and related product wholesalers	X		
4451	Grocery stores		X	
4481	Clothing stores			X
4511	Sporting goods and musical instrument stores	X		X
4541	Electronic shopping and mail-order houses	X		
5111	Newspaper, book, and directory publishing	X	X	
5112	Software publishers	X		X
5121	Motion picture and video industries	X	X	
5151	Radio and television broadcasting		X	
5152	Cable and other subscription programming	X		
5174	Satellite telecommunications	X		
5179	Other telecommunications	X		
5191	Other information services	X		
5221	Depository credit intermediation	X		
5239	Other financial investment activities	X		
5241	Insurance carriers	X	X	
5311	Lessors of real estate	X		
5331	Lessors of nonfinancial intangible assets	X		
5416	Management and technical consulting	X		X
5417	Scientific research and development	X		X
5614	Business support services	X		
5615	Travel arrangement and reservation	X		
6214	Outpatient care centers	X		
7132	Gambling industries	X	X	

Source: ESA and USPTO calculations.

Brands

Firms with high-value brands are likely to seek protection for those brands through trademark law. To check robustness of the methodology for selected trademark-intensive industries, we examined the Interbrand listing of the "Best Global Brands in 2010," focusing on the extent to which firms with highly valued corporate brands operate in industries that have been identified as trademark-intensive.[51]

Methodology

Briefly, Interbrand's methodology for ranking global brands considers the ongoing investment and management of the brand as a business asset.[52] Three key aspects contribute to the assessment of each brand:

- Financial performance measures an organization's raw financial return to the investors.
- Role of brand measures the portion of the decision to purchase that is attributable to brand.
- Brand strength measures the ability of the brand to secure the delivery of expected future earnings.

Before discussing the brand rankings, it is important to review caveats and differences in scope between Interbrand's approach and our approach for identifying trademark-intensive industries. To be included in Interbrand's ranking, companies must have corporate brands that are truly "global, visible, and relatively transparent." First, the company must be present on at least three continents, including "broad coverage in growing and emerging markets." At least 30 percent of revenues must be from outside of the home country, with no more than 50 percent from any single continent. Second, the brand's public profile must extend beyond its own marketplace. Third, firms must prove their profitability through publicly available financial data. These criteria are restrictive relative to the rest of this report, in which the general focus is on all corporations, regardless of their global presence or profits.

Also, it should be emphasized that "brand" refers to single corporate brands as opposed to product brands. This distinction led Interbrand to exclude pharmaceutical companies, which are considered better identified through specific products than the companies themselves. Similarly, Wal-Mart is excluded not because it is insufficiently global as a company, but because it does not operate under a single Wal-Mart brand globally.[53]

The corporate brands included in Interbrand's Best Global Brands are identified by company name, and we assigned four-digit NAICS codes to each listed company using OneSource, following the same methodology we used to assign NAICS codes to the top 50 trademark registrants.[54] We then compared the industries with top-ranked brands with the industries selected as trademark-intensive.

Results

We discovered considerable overlap between the trademark-intensive industries and those with high-ranking brands. Upon assigning NAICS codes to each brand, we calculated that the top 100 ranked brands corresponded to 41 different four-digit NAICS industries, and 24 of these industries (covering 70 brands) were selected as trademark-intensive. Put another way, 24 of the 60 trademark-intensive industries also had high-ranking brands either in the United States or abroad.

Table 7 lists the 24 trademark-intensive industries in which 70 of the various top 100 brands operate.[55] Beverage manufacturing (NAICS 3121) was the most heavily represented industry on the list, with 11 of the top 100 brands—including Coca-Cola, Interbrand's #1 ranked global brand, as well as Pepsi (#23), Budweiser (#30), Sprite (#61), Jack Daniel's (#80), and six foreign brands.[56] This industry is noteworthy because it also was the only one to be selected as trademark-intensive based on all three measures.

While there is much overlap between the trademark-intensive industries and the top-ranked brands, there also were 17 industries that were not considered trademark-intensive based on any of our three measures. Several happened to be among the top 10 brands, including IBM (#2, NAICS 5415), McDonald's (#4, NAICS 7222), General Electric (#5, NAICS 5222), and Intel (#7, NAICS 3344). Nonetheless, these brands are captured to some extent elsewhere in the report. IBM and Intel operate in industries that are found to be copyright- and patent-intensive industries, respectively. General Electric, like a number of the companies examined in this report, operates in various industries. Although its consumer finance arm is in an industry that is not found to be trademark-intensive, other parts of the company are in trademark-intensive industries, such as turbine and power transmission equipment (NAICS 3336). The food service industry, such as limited-service eating places (NAICS 7222), is not captured by any of the trademark-intensive measures or elsewhere in this report. This is not to say that trademarks or other forms of IP protection are unimportant for protecting food-service industry brands, four of which are in the top 100.

Trademark Registrations by Class

As discussed earlier, the USPTO organizes trademark registrations by class as opposed to industry. For several reasons, it is not possible to use the registration class data to identify trademark-intensive industries by four-digit

NAICS industry. First, the classes are not sufficiently detailed for our purposes.

There are 49 trademark classes, as opposed to 313 four-digit NAICS industries. So, even if it were possible to clearly define the correspondence between class and NAICS, a meaningful level of detail would be lacking.

Second, there is no straightforward correspondence between trademark classes and industries because the classes and NAICS industries are conceptually quite distinct.

NAICS industries are defined according to the principal business activity of an establishment. Trademark classes indicate the type of good or service associated with a particular trademark. In a sense, the classes tell us what type of product is carrying the protected mark or logo. Consider the Whole Foods example cited earlier.

Their "365 Everyday Value" logo appears on a wide range of goods covering 10 different trademark classes. As another example, consider Five Guys, the fast food franchise.

This company has protected its "Five Guys" logo within six different trademark classes, ranging from meats and processed foods (for the sandwiches), to staple foods (processed peanut and fried potatoes), non-alcoholic beverages, advertising and business (franchising), clothing, and vehicles (to protect license plate holders with the logo).[57]

Within the NAICS framework, however, Five Guys would simply be considered a limited-service eating place (NAICS 7222), a service-providing industry. So, while Five Guys is a service-provider under NAICS, most of its trademark classes relate to goods, that is, the "vessels" on which the company logo appears.

Despite the difficulty in precisely relating trademark classes to NAICS industries, it is worthwhile to look for similarities between the distribution of trademark registrations by class and our selection of trademark-intensive industries.

Table 8 lists total registrations by trademark class for the period from November 16, 1999 to November 15, 2009, ranking the registration classes by trademark count. These data provide insight into which areas of the economy (by class) take advantage most often of trademark protection through the Federal trademark registration system.

The top seven categories accounted for more than half of all registrations in the 2000s. Three of these categories—advertising and business, education and entertainment, and insurance and financials—are very broad and cover services ranging from wholesale and retail trade to professional and business services; financial services; insurance; educational services; and the arts, entertainment, and recreation industry.

Table 7. Trademark-Intensive Industries with Top 100 Global Brands in 2011

NAICS code	Industry title	Selection criteria		
		Trademark intensity	Top 50	Sample
3371	Household and institutional furniture	X		
3391	Medical equipment and supplies manufacturing	X		
3399	Other miscellaneous manufacturing	X	X	
4234	Commercial equip. merchant wholesalers	X		X
4242	Druggists' goods merchant wholesalers	X		
4244	Grocery and related product wholesalers	X		
4451	Grocery stores		X	
4481	Clothing stores			X
4511	Sporting goods and musical instrument stores	X		X
4541	Electronic shopping and mail-order houses	X		
5111	Newspaper, book, and directory publishing	X	X	
5112	Software publishers	X		X
5121	Motion picture and video industries	X	X	
5151	Radio and television broadcasting		X	
5152	Cable and other subscription programming	X		
5174	Satellite telecommunications	X		
5179	Other telecommunications	X		
5191	Other information services	X		
5221	Depository credit intermediation	X		
5239	Other financial investment activities	X		
5241	Insurance carriers	X	X	
5311	Lessors of real estate	X		
5331	Lessors of nonfinancial intangible assets	X		
5416	Management and technical consulting	X		X
5417	Scientific research and development	X		X
5614	Business support services	X		
5615	Travel arrangement and reservation	X		
6214	Outpatient care centers	X		
7132	Gambling industries	X	X	

Source: ESA calculations using Interbrand's Top 100 Global Brands and data from OneSource.

At a superficial level, it rings true that these sectors rely on trademarks because they cover many of the services and brands that consumers use on a daily basis.

Not surprisingly, several of the trademark-intensive industries correspond to these classes; however, many other industries do as well.

On the other hand, consider the trademark classes for light beverages as well as wine and spirits, both of which would be expected to correspond to beverage manufacturing (NAICS 3121). This industry is clearly trademark-intensive as it was the only industry to be selected by all three criteria (trademark intensities, top 50 list, and sample).

Furthermore, several beverages appeared in the top 100 brands Yet, the light beverages class only accounted for 0.8 percent of registrations (15,253) while wine and spirits accounted for 1.1 percent (20,727) for FY 2000 to FY 2009.

Copyrights

Overview

Copyrights, the third form of IP rights covered in this report, protect "original works of authorship."[58] These works must be fixed in a tangible form of expression, meaning that concepts that never leave the confines of our minds cannot be copyrighted. Protection under copyright, which lasts for the life of the author plus an additional 70 years, is secured automatically when a work is created.[59]

Neither publication nor registration with the Copyright Office is required to secure copyright protection. But registering a copyright does establish a public record of the copyright, and it can be beneficial because of incentives provided to encourage registration. For example, a registered copyright can be recorded with the U.S. Customs and Border Protection to protect against the importation of infringing copies.[60]

More than 33.7 million copyrights have been registered in the United States since 1790, when Congress enacted the first Federal copyright law. Of these, approximately 150,000 were registered between 1790 and the centralization of copyright functions in the Library of Congress in 1870.[61] In 2009, more than 382,000 basic copyrights were registered.

Table 8. Trademark Registrations by Class, Ranked by Number of Registrations, FY 2000-09

Trademark class	Class title	Trademark registrations		
		Total	Percent of total	Cumulative percent
9	Electrical and scientific apparatus	209,639	11.6%	11.6%
35	Advertising and business	184,274	10.2%	21.9%
41	Education and entertainment	151,547	8.4%	30.3%
42	Computer and scientific	136,025	7.5%	37.8%
16	Paper goods and printed matter	97,739	5.4%	43.2%
36	Insurance and financial	92,759	5.1%	48.4%
25	Clothing	91,923	5.1%	53.5%
5	Pharmaceuticals	57,081	3.2%	56.6%
28	Toys and sporting goods	53,292	3.0%	59.6%
3	Cosmetics and cleaning preparations	51,940	2.9%	62.5%
30	Staple foods	43,550	2.4%	64.9%
37	Building construction and repair	39,557	2.2%	67.1%
7	Machinery	36,683	2.0%	69.1%
38	Telecommunications	32,482	1.8%	70.9%
11	Environmental control apparatus	31,374	1.7%	72.7%
10	Medical apparatus	30,872	1.7%	74.4%
20	Furniture and articles, nec.	29,527	1.6%	76.0%
29	Meats and processed foods	27,188	1.5%	77.5%
1	Chemicals	27,009	1.5%	79.0%
44	Medical, beauty, and agricultural	26,404	1.5%	80.5%
12	Vehicles	25,891	1.4%	81.9%
43	Hotels and restaurants	25,825	1.4%	83.4%
39	Transportation and storage	25,562	1.4%	84.8%
21	Housewares and glass	25,356	1.4%	86.2%
6	Metal goods	21,724	1.2%	87.4%
14	Jewelry	21,422	1.2%	88.6%
18	Leather goods	20,843	1.2%	89.7%
33	Wine and spirits	20,727	1.1%	90.9%
40	Treatment of materials	19,532	1.1%	92.0%
19	Nonmetallic building materials	17,801	1.0%	93.0%
32	Light beverages	15,253	0.8%	93.8%
31	Natural agricultural products	14,744	0.8%	94.6%
45	Personal	14,390	0.8%	95.4%
24	Fabrics	13,233	0.7%	96.2%
17	Rubber goods	11,637	0.6%	96.8%
8	Hand tools	11,098	0.6%	97.4%
2	Paints	8,352	0.5%	97.9%
4	Lubricants and fuels	7,502	0.4%	98.3%

Table 8. (Continued)

Trademark class	Class title	Trademark registrations		
		Total	Percent of total	Cumulative percent
18	Leather goods	20,843	1.2%	89.7%
33	Wine and spirits	20,727	1.1%	90.9%
40	Treatment of materials	19,532	1.1%	92.0%
19	Nonmetallic building materials	17,801	1.0%	93.0%
32	Light beverages	15,253	0.8%	93.8%
31	Natural agricultural products	14,744	0.8%	94.6%
45	Personal	14,390	0.8%	95.4%
24	Fabrics	13,233	0.7%	96.2%
17	Rubber goods	11,637	0.6%	96.8%
8	Hand tools	11,098	0.6%	97.4%
2	Paints	8,352	0.5%	97.9%
4	Lubricants and fuels	7,502	0.4%	98.3%
34	Smokers' articles	5,004	0.3%	98.6%
27	Floor coverings	4,771	0.3%	98.8%
26	Fancy goods	4,698	0.3%	99.1%
22	Cordage and fibers	3,809	0.2%	99.3%
13	Firearms	3,628	0.2%	99.5%
15	Musical instruments	3,555	0.2%	99.7%
200	Collective membership	1,577	0.1%	99.8%
23	Yarns and threads	1,423	0.1%	99.9%
B	Services certification mark	1,391	0.1%	100.0%
A	Goods certification mark	765	0.0%	100.0%
	Total	1,802,378	100.0%	100.0%

Source: ESA calculations using unpublished USPTO data.

Note: The cumulative percent figures may not equal the sum of the percent of total figures because of rounding.

The types of works that can be copyrighted are outlined in the 1976 Copyright Act. A partial list of copyrighted materials includes the following:

- literary works (including fiction, nonfiction, and computer programs)
- musical works, including any accompanying words
- dramatic works, including any accompanying music
- pantomimes and choreographic works
- pictorial, graphic, and sculptural works
- motion pictures and other audiovisual works
- sound recordings
- architectural works, including vessel hull designs

Of the 382,000 registrations in 2009, about 93,000 covered performing arts works, such as films, musical and dramatic works, and choreography. There were 75,000 registrations of works of visual and applied arts. These covered fine and graphic arts, sculptures, technical drawings and models, photographs and cartographic works, and commercial prints and labels. Registrations of sound recordings totaled 42,000.[62]

Identifying Copyright-Intensive Industries

Our methodology for designating copyright-intensive industries draws heavily from definitions established by the World Intellectual Property Organization's (WIPO) *Guide on Surveying the Economic Contribution of the Copyright-based Industries.*[63] A series of reports by Stephen Siwek entitled *Copyright Industries in the U.S. Economy* have applied these definitions to the U.S. economy.[64] While this established literature underlies our analysis, we used a more narrow definition of copyright-intensive industries than WIPO, focusing on industries that produce copyrighted work and excluding several industries associated with the distribution of copyrighted material. This deviation from the WIPO Guide was needed in order to maintain internal consistency with our measures of patent- and trademark-intensive industries.

Methodology

Because WIPO's *Guide on Surveying the Economic Contribution of the Copyright-based Industries* clearly distinguishes the type of works that can be copyrighted, the industries in which those works are created, and the downstream (distribution) industries delivering the produced copyrighted works, it is possible to develop a list of copyright-intensive industries that is comparable in scope to our lists of patent- and trademark-intensive industries. We started by focusing on "core" copyright industries, which WIPO defines as industries "wholly engaged in creation, production and manufacturing, performance, broadcast, communication and exhibition, or distribution and sales of works and other protected subject matter."[65] In other words, core copyright industries are considered 'core' because they either produce copyrighted materials or bring them to market.

For this report, we are only concerned with the set of industries that are primarily responsible for the *creation or production* of copyrighted materials and designate them as copyright-intensive. Thus, to the extent possible using four-digit NAICS industry codes, we excluded industries whose primary purpose is to distribute copyright materials to businesses, consumers, or both. For example, we did not count industries such as book, periodical, and music stores (NAICS 4512) or consumer goods rental (NAICS 5322), which includes video rentals, as copyright-intensive even though they are part of the "core" category in the WIPO guide.[66] Our definition is narrower than WIPO's in order to be consistent with our treatment of patent- and trademark-intensive industries, where the industries most responsible for the production of protected IP are the main focus. This approach simply reflects our goal of examining the industries in the economy that are most responsible for the production of protected IP and use of this approach is not a criticism of WIPO's guidelines.

One conceptual distinction between the approach taken here and that selected in the patent and trademark sections is worth noting. Throughout this report, the focus was on industries that produce protected IP, whether patents, trademarks, or copyrights. In the case of patents and trademarks, we defined "intensive" industries as the subset of all patent or trademark producers that had high scores in various "intensity" measures, whereas we defined as copyright-intensive essentially all industries associated with the production of copyrighted materials.

Results

Table 9 lists the copyright-intensive industries. All are involved in the creation and/or recording (in print, magnetically, or digitally) of protected works. For industries such as other information services (NAICS 5191), the title of which does not clearly indicate what protected materials are produced, the relevant copyrighted product is listed in parentheses.

Some of the selected industries are involved in both the production and distribution of copyrighted materials since both functions are often performed within a single business establishment.[67] The newspaper industry provides a good example of the blurry line between the creation and distribution of copyrighted materials. We considered newspaper, periodical, book and directory publishers (NAICS 5111) to be copyright-intensive because much copyrighted material actually is developed within establishments classified in this industry, as evidenced by the fact that this industry had 31,000 reporters and correspondents, 25,000 graphic designers, and 4,000 photographers on its payrolls in 2010.[68]

This line of reasoning underlies the inclusion of several industries on the list, such as sound recording, broadcasting, cable programming, and performing arts companies.

Table 9. Copyright-Intensive Industries

NAICS code	Industry title
5111	Newspaper, periodical, book, and directory publishers
5112	Software publishers
5121	Motion picture and video industries
5122	Sound recording industries
5151	Radio and television broadcasting
5152	Cable and other subscription programming
5191	Other information services (news syndicates and internet sites)
5414	Specialized design services (visual and graphic arts)
5415	Computer systems design and related services (software and databases)
5418	Advertising, public relations, and related services
5419	Other professional, scientific, and technical services (photography and translation)
7111	Performing arts companies
7115	Independent artists, writers, and performers

Source: ESA selection based on World Intellectual Property Organization 2003.

These industries have many "creators" (writers, composers, choreographers, and others) directly on their payrolls, in addition to other independent artists and content creators under individual contracts with their companies.

Our list of copyright-intensive industries also includes two additional industries directly involved in the writing and publishing of newspapers: independent artists, writers, and performers (NAICS 7115) and other information services (NAICS 5191), which includes news syndicates. However, we excluded printing and related support activities (NAICS 3231), because its role is simply to print the newspapers themselves. WIPO considers this a "core" copyright industry because it is directly involved in creating a "vessel" for copyrighted material, meaning that it exists in order to deliver copyrighted materials in a tangible form to a consumer or audience. However, printing is distinct from the actual development of the copyrighted materials (such as news articles and photographs), so we did not consider it to be a copyright-intensive industry.

As another example, consider the manufacturing and reproducing magnetic and optical media industry (NAICS 3346). This industry produces software CDs and DVDs, as well as other prerecorded disks, tapes and vinyl records. While inextricably linked to copyright-intensive industries, this industry is not part of the actual process of developing and recording software or music, but rather a part of the downstream process of putting the material on a tangible vessel. So as with the printing example above, while manufacturing and reproducing magnetic and optical media is a WIPO-defined core industry, it is not counted as copyright-intensive in this report. It should be noted, however, that both printing and related support activities (NAICS 3231) and manufacturing and reproducing magnetic and optical media (NAICS 3346) were found to be trademark-intensive.

IP-Intensity and the Use of Intellectual Property Protection

Before reviewing the complete list of IP-intensive industries, we should note that defining "IP intensity" as the use of intellectual property protection is an approach open to criticism. Although commonly used in the scientific literature, such intensity measures tend to overweight those industries that more commonly use patents, trademarks, or copyrights for purposes other than activities closely related to R&D or commercialization *per se*.[69]

The term "IP-intensive" defined in this report is not necessarily directly related to the value or purpose of the IP held by companies. In some industries a single patent may support revenues in the billions of dollars, while in other industries, many patents may be required just to market marginally profitable products.[70,71] The same logic holds for trademarks and copyrights.[72]

For example, the Carnegie-Mellon University survey, previously mentioned, shows significant differences across industries in how companies acquire and use patents.[73] These findings are not surprising. First, companies employ other protections, such as trade secrets, that may be a substitute for patent protection. And, because the term of protection is limited, patent efficacy can be diminished by lengthy product and process development. Often companies will seek patent protection early in that timeline (as patent costs remain relatively inexpensive in the United States), only to determine after further research and development that the patent protection was neither needed nor particularly useful in the final outcome.

Prior economic studies have also shown that patents can be used for defensive purposes, and our method for selecting IP-intensive industries necessarily sweeps widely, including industries in which companies seek patent protection regardless of the purpose.[74]

To understand the various motivations for patenting, the Carnegie-Mellon study also focused on the reasons companies in different industries sought patent protection and the multiple uses of patents. The reasons included earning licensing revenue, use in negotiations, blocking competitors, enhancing reputation, or as a measure of employee performance.

The research found that companies reported that their patents often served multiple purposes.

All drug companies surveyed, for instance, used product patenting to prevent copying, while 97 percent used product patenting to block competitors, 69 percent to enhance the company's reputation, and 61 percent to aid in negotiations. In the aerospace industry, 57 percent of companies reported using product patents for licensing revenue, while 50 percent of those companies used process patents for the same purpose. Among companies in the machine tool sector, 13 percent reported that they used product patents as a measure of employees' performance.

Collectively, these findings show that the term "use" is complex at the company level, and that complexity must be acknowledged in interpreting the aggregated measures we report at the industry level.

IP-Intensive Industries

Seventy-five IP-intensive industries emerged after combing through the lists of patent-, trademark-, and copyright-intensive industries. (See Table 10.) This 75-industry total is smaller than the sum of the parts. Indeed, there is considerable overlap between the patent-intensive and trademark-intensive industry lists, as 18 of the 26 patent-intensive industries also were identified among the 60 trademark-intensive ones. There also is some overlap between the copyright-intensive and trademark-intensive list with six of the 13 copyright-intensive industries also selected as trademark-intensive. By definition, however, there is no overlap between the patent and copyright lists. The USPTO classified all patents within manufacturing industries, whereas industries involved in the creation and/or recording of protected works are classified as service-providing industries and not in the manufacturing sector.

Table 10 also lists total employment of each IP-intensive industry. Employment refers to wage and salary jobs as well as self-employment, which accounts for a sizeable portion of employment in the copyright-intensive industries.[75] Employment in IP-intensive industries totaled 27.1 million jobs in 2010. The next section will examine in more detail their contribution to overall employment and gross domestic product in the U.S. economy as well as workers' wages, educational attainment, and foreign trade.

Table 10. IP-Intensive Industries and Selection Criteria

NAICS code	Industry title	Selection criteria			
		Employment in 2010 (1000 jobs)	Patent-intensive	Trademark-intensive	Copyright-intensive
2111	Oil and gas extraction	163.3		X	
2361	Residential building construction	934.3		X	
3112	Grain and oilseed milling	58.3		X	
3113	Sugar and confectionery product manufacturing	69.9		X	
3115	Dairy product manufacturing	130.4		X	
3119	Other food manufacturing	165.5		X	
3121	Beverage manufacturing	169.5		X	
3162	Footwear manufacturing	13.5		X	
3221	Pulp, paper, and paperboard mills	112.7		X	
3222	Converted paper product manufacturing	285.9		X	
3231	Printing and related support activities	524.4		X	
3251	Basic chemical manufacturing	143.1	X	X	
3252	Resin, rubber, and artificial fibers	90.1	X	X	
3253	Agricultural chemical manufacturing	36.7	X		
3254	Pharmaceutical and medicine manufacturing	277.6	X	X	
3255	Paint, coating, and adhesive manufacturing	56.4	X		
3256	Soap, cleaning compound, and toiletries	104.2	X	X	
3259	Other chemical product and preparations	84.9	X	X	
3261	Plastics product manufacturing	501.5		X	
3322	Cutlery and handtool manufacturing	42.1		X	
3329	Other fabricated metal product manufacturing	256.4		X	
3331	Ag., construction, and mining machinery manufacturing	208.9	X		
3332	Industrial machinery manufacturing	102.2	X	X	
3333	Commercial and service industry manufacturing	91.3	X	X	
3334	HVAC and commercial refrigeration	124.4	X		
3335	Metalworking machinery manufacturing	156.1	X		
3336	Turbine and power transmission equipment manufacturing	91.5	X	X	
3339	Other general purpose machinery manufacturing	229.8	X	X	
3341	Computer and peripheral equipment	161.9	X	X	
3342	Communications equipment manufacturing	118.5	X	X	
3343	Audio and video equipment manufacturing	20.2	X	X	

3344	Semiconductor and electronic component manufacturing	373.8	X		
3345	Electronic instrument manufacturing	408.7	X	X	
3346	Magnetic media manufacturing and reproducing	26.3	X		
3351	Electric lighting equipment manufacturing	46.2	X		
3352	Household appliance manufacturing	60.7	X	X	
3353	Electrical equipment manufacturing	137.3	X	X	
3359	Other electrical equipment and components	118.7	X	X	
3361	Motor vehicle manufacturing	151.8		X	
3371	Household and institutional furniture	245.8		X	
3391	Medical equipment and supplies manufacturing	311.5	X	X	
3399	Other miscellaneous manufacturing	309.6	X	X	
4234	Commercial equip merchant wholesalers	615.4		X	
4242	Druggists' goods merchant wholesalers	193.3		X	
4244	Grocery and related product wholesalers	735.3		X	
4451	Grocery stores	2,521.6		X	
4481	Clothing stores	1,114.4		X	
4511	Sporting goods and musical instrument stores	514.6		X	
4541	Electronic shopping and mail-order houses	303.9		X	
5111	Newspaper, book, and directory pub	530.9		X	X
5112	Software publishers	259.8		X	X
5121	Motion picture and video industries	414.6		X	X
5122	Sound recording industries	36.4			X
5151	Radio and television broadcasting	216.7		X	X
5152	Cable and other subscription programming	85.6		X	X
5174	Satellite telecommunications	14.6		X	
5179	Other telecommunications	127.5		X	
5191	Other information services	145.5		X	X
5221	Depository credit intermediation	1,735.6		X	
5239	Other financial investment activities	427.4		X	
5241	Insurance carriers	1,368.2		X	
5311	Lessors of real estate	823.9		X	
5331	Lessors of nonfinancial intangible	25.5		X	

Table 10. (Continued)

NAICS code	Industry title	Employment in 2010 (1000 jobs)	Selection criteria Patent-intensive	Trademark-intensive	Copyright-intensive
5414	Specialized design services	242.6			X
5415	Computer systems design and related services	1,595.1			X
5416	Management and technical consulting	1,233.8		X	
5417	Scientific research and development	638.2		X	
5418	Advertising, PR, and related services	461.2			X
5419	Other professional and technical services	673.4			X
5614	Business support services	883.9		X	
5615	Travel arrangement and reservation	201.7		X	
6214	Outpatient care centers	636.6		X	
7111	Performing arts companies	132.1			X
7115	Independent artists, writers, and performers	306.1			X
7132	Gambling industries	133.9		X	

Source: ESA calculations using data from the Bureau of Labor Statistics' Industry Productivity program.

Note: Employment includes wage and salary jobs, the self-employed, and unpaid family workers and is measured in thousands of jobs.

III. IP-Intensive Industries in the Economy

Employment

Employment totaled 27.1 million jobs in IP-intensive industries in 2010, representing 18.8 percent of all jobs in the economy. (See Figure 1.) Our definition of jobs includes not just payroll (or wage and salary workers), but also the self-employed and unpaid family workers.[76] Labor market analyses of this type often overlook the self-employed, but as will be seen, they are a critical source of staff for copyright-intensive industries, and to a lesser extent, for trademark-intensive industries.

Given that 60 of the 75 IP-intensive industries were considered trademark-intensive, it follows that these industries would account for the majority of IP-intensive jobs (22.6 million or 83 percent of all IP-intensive jobs). Indeed, employment in the patent- and copyright intensive industries was significantly lower. In 2010, patent-intensive industries had 3.9 million jobs and copyright-intensive industries had 5.1 million jobs. The 24 industries that were considered intensive with respect to more than one form of IP protection had 4.5 million jobs.[77]

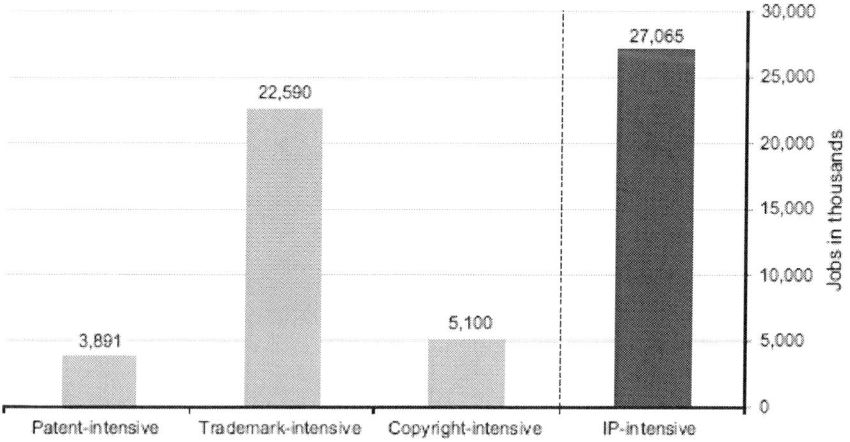

Source: ESA calculations using data from the Bureau of Labor Statistics' Industry Productivity program.

Note: Estimates include wage and salary employment, the self-employed, and unpaid family workers. Because several industries were found to be trademark-intensive and patent- or copyright-intensive, total employment in IP-intensive industries is less than the sum of employment in patent-, trademark-, and copyright-intensive industries.

Figure 1. Employment in IP-Intensive Industries, 2010.

As can be seen in Figure 2, employment growth over the past two decades has varied considerably across IP-intensive industries. The figure compares the relative job growth in the various industry groupings since 1990. Employment in IP-intensive industries in 2010 was little changed from 1990 as the 12-percent growth in employment during the 1990s was subsequently reversed during the 2000s. In contrast, non-IP-intensive employment grew in the 1990s and had relatively flat employment in the 2000s.[78] As a result, the IP-intensive industries' share of total employment edged down from 21.7 percent in 1990 to 20.7 percent in 2000 and declined nearly 2 more percentage points by 2010 to 18.8 percent.

Overall employment in IP-intensive industries has lagged other industries over the past two decades due primarily to historic losses in manufacturing jobs. This decline tracks with other statistics showing that during the last decade the economy was underinvesting and underperforming in manufacturing. For much of the 2000s, manufacturing capacity remained stagnant, and for the first time in decades, manufacturing firms' real investment in fixed assetsstagnated.79 During this same period, the U.S. manufacturing sector lost more than 3 million jobs.[80]

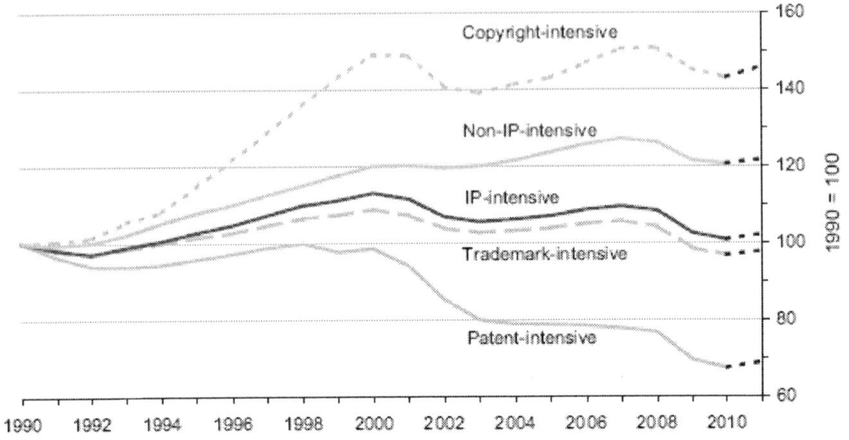

Source: ESA calculations using data from the Bureau of Labor Statistics' Industry Productivity program.

Note: Black dotted lines represent ESA projections of 2011 employment growth based on data from the Bureau of Labor Statistics' Current Employment Statistics program.

Figure 2. Indexed Employment in IP-Intensive Industries, 1990-2011.

Over the two decades, shrinking employment in patent-intensive industries was offset by expansion in the copyright-intensive ones. Patent-intensive industries are a subset of the manufacturing sector, and its job losses are much like those experienced throughout the nation's manufacturing sector. The copyright-intensive industries largely belong to the information and professional and technical services industry sectors in NAICS, and their employment trends parallel the trends seen in those sectors: steady growth in the 1990s and offsetting gains and losses in the 2000s.[81] That is, job growth between 2000 and 2010 in professional and technical services was largely offset by employment declines in the information industries.

As the economic recovery has unfolded, real business fixed investment in equipment and software has rebounded, growing by nearly 33 percent from the second quarter of 2009 through the end of 2011.[82] At the same time, manufacturing production surged at a 5.7 percent annual rate from June 2009 through December 2011, the fastest pace of growth of production in a decade.[83]

The resurgence in production led manufacturers to add 346,000 factory jobs in 2010 and 2011, the strongest two-year period of manufacturing job growth since the late 1990s.[84]

Preliminary employment projections from the Economics and Statistics Administration for 2011 show that as the economic recovery has taken hold, IP-intensive industry employment has rebounded and contributed disproportionately to overall employment growth. As shown in Figure 2, job growth was widespread, with a 1.6 percent rise in employment in IP-intensive industries outpacing the 1.0 percent increase in non-IP-intensive industries. Breaking IPintensive industries out into its constituent parts uncovers 2.4 percent job growth in copyright-intensive industries, 2.3 percent growth in patent-intensive industries, and 1.1 percent growth in trademark-intensive industries.

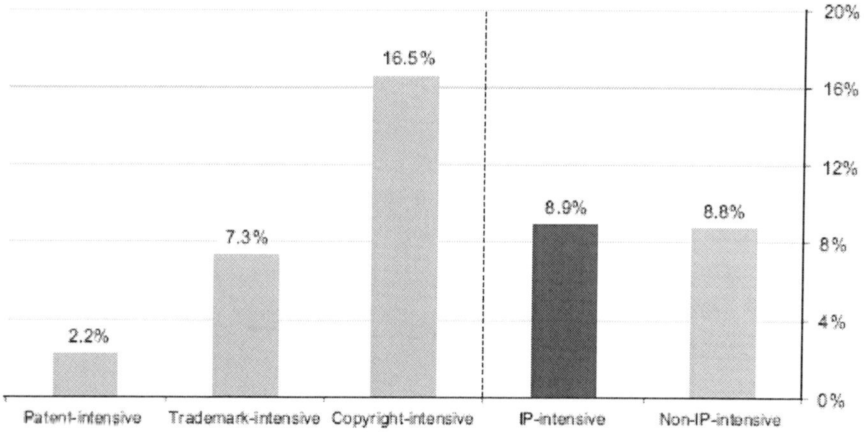

Source: ESA calculations using data from the Bureau of Labor Statistics' Industry Productivity program.

Note: Estimates show the self-employed and unpaid family workers as a share of all jobs. However, unpaid family workers account for only about one percent of the combined total of the self-employed and unpaid family workers outside of agriculture.

Figure 3. Self-Employed Share of All Jobs in IP-Intensive Industries, 2010.

In 2010, the self-employed filled 2.4 million jobs in IP-intensive industries. This 8.9 percent self-employment share was essentially equal to the 8.8 percent share in other (non-IP-intensive) industries; however, there was notable variation across IP-intensive industries (See Figure 3). Trademark-intensive industries had the largest number of self-employed persons at 1.7

million or 7.3 percent of all trademark-intensive jobs. The highest self-employment share, however, was in the copyright-intensive industries, in which the 0.8 million self-employed workers filled 16.5 percent of all jobs. This high share is not surprising as many jobs in the creative and performing arts are contract rather than payroll jobs, usually related to the completion or performance of a specific authored work.

Total Employment Supported by IP-Intensive Industries

While IP-intensive industries had 27.1 million jobs either on their payrolls or under contract in 2010, these industries indirectly supported an additional 12.9 million jobs in other (non-IPintensive) industries throughout the economy. In other words, IP-intensive industries supported 40.0 million jobs (or 27.7 percent of all jobs) directly and indirectly, through a supply chain that stretches across the economy.[85]

Figure 4 expands on Figure 1, showing employment in IP-intensive industries (medium blue bar) plus indirect employment in the supply chain (the rectangle below the blue bar). The figure also shows the jobs that the patent-, trademark-, and copyright-intensive industries support in their respective (non-patent-, trademark-, and copyright-intensive) supply chains. Thus, as shown, patent-intensive industries supported an additional 3.3 million workers indirectly. Likewise, trademark-intensive industries supported 13.1 million jobs indirectly, and copyright-intensive industries indirectly via the supply chain supported 2.5 million jobs.

As these data suggest, patent-intensive industries relied more deeply than the trademark- and copyright-intensive industries on an outside supply chain and indirectly supported a relatively large number of jobs. Consider that the patent-intensive industries, despite being smaller in terms of employment than the copyright-intensive ones, supported a larger multiple of outside supply chain employment. This is because patent-intensive industries come from the manufacturing sector, which typically has a larger multiplier effect than the service sector, from which the copyright-intensive industries are drawn.

Additionally, as the "intensive" industry lists become longer, the number of "non-intensive" industries, which is the source of the indirect, supply chain jobs, becomes shorter.

This is why trademark-intensive industries alone indirectly supported 13.1 million workers, while IPintensive industries as a whole indirectly supported only 12.9 million workers.

This is also why the 60 trademark-intensive industries were more likely to self-supply some of their inputs than the 26 patent-intensive industries and less likely to rely on outside industries.

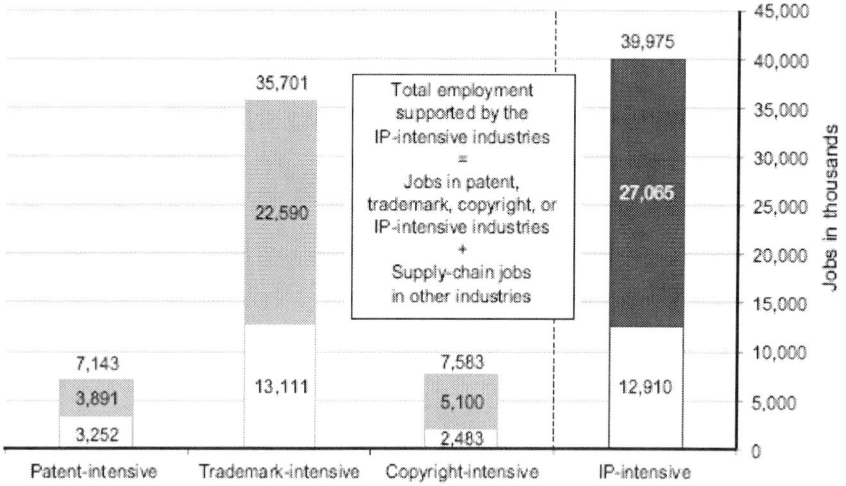

Source: ESA calculations using data from the Bureau of Economic Analysis' Industry Accounts office and the Bureau of Labor Statistics' Industry Productivity program.
Note: Estimates include wage and salary employment, the self-employed, and unpaid family workers.

Figure 4. Total Employment Supported by IP-Intensive Industries, 2010.

Put another way, the potential outside supply chain (indirect jobs) for patent-intensive industries was larger (covers more industries) than the potential supply chain for the trademark-intensive industries.

These results may appear conservative relative to multiplier analyses often cited in the economic literature with employment multipliers of about 2. Such multipliers tell us that if growing demand for final goods and services (as opposed to intermediate inputs to production) spurs companies to add 100 new jobs, then an additional 100 jobs will be added indirectly throughout the economy.[86]

We estimated that for IP-intensive industries this multiplier would be about 1.8, which is close to the multiplier of 2 found in similar industry analyses.

As noted earlier, because a relatively large number of industries are considered IP-intensive, much of the IP supply chain is internal to these industries, which reduces the multiplier effect relative to analyses that focus on a single industry. At the extreme, the jobs-to-jobs multiplier could be calculated for the entire economy and would be 1 because there would be no external supply chain left.

In this report, the question that is being addressed differs subtly from the question that can be answered by a traditional multiplier analysis. Rather than focusing on a hypothetical change in final demand for IP-intensive goods and services, we were interested in learning which IPintensive jobs were tied to final demand, which were in the supply chain, and how many jobs in other industries were part of that supply chain.

Teasing out these data from the input/output framework, we estimated that 16.2 million jobs in IP-intensive industries were associated with producing goods and services to satisfy final demand while 10.9 million jobs in these industries were associated with production for the supply chain.

An additional 12.9 jobs in other (nonIP-intensive industries) also were part of the IP supply chain.[87]

As is typical with input/output analyses, these data emphasize upstream industries (supply chain) that provide intermediate inputs needed to produce goods and services in IP-intensive industries. They do not, however, capture the significant downstream channels that facilitate the distribution and trade in IP-intensive goods and services.

Therefore, the estimate that there were 40.0 million jobs supported by IP-intensive industries is actually a conservative estimate of employment linked to these industries.

Value Added

While IP-intensive industries accounted for 18.8 percent of all jobs in the economy in 2010, their $5.06 trillion in value added in 2010 represented 34.8 percent of total GDP.[88] This total share of GDP has edged down since 2003. Because 60 of the 75 IP-intensive industries were considered trademark-intensive, it is unsurprising that this segment alone accounted for almost 31 percent of GDP with $4.5 trillion in value added in 2010. (See Figure 5.) Patent-intensive and copyright-intensive industries accounted for 5.3 and 4.4 percent of GDP, with $763 billion and $641 billion in value added, respectively.

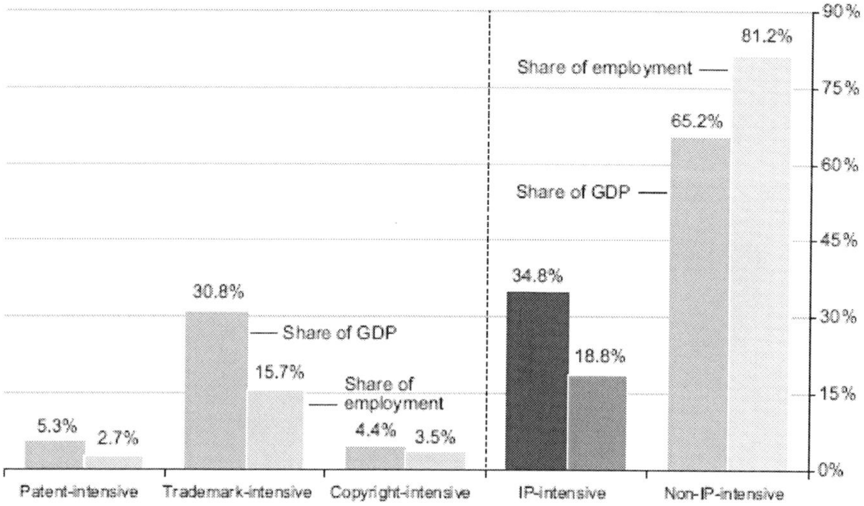

Source: ESA calculations using data from the Bureau of Economic Analysis, National Income and Product Accounts and the Bureau of Labor Statistics' Industry Productivity program.

Figure 5. Value Added and Employment Shares of IP-Intensive Industries, 2010.

Payroll Employment by State in IP-Intensive Industries

The IP-intensive share of all covered employment varies notably across states and regions, as shown in Map 1.[89] Four of the top five states in IP-intensive employment share were in New England: Massachusetts (23.6 percent), New Hampshire (22.0 percent), Connecticut (21.7 percent), and Vermont (21.6 percent).

Wisconsin, with a 23.0 percent share, occupied second place on the list. Overall there were 16 states above the national average of 19.1 percent.

These states were spread throughout the mid-Atlantic, Northeast, and Midwest and also include California, Colorado, and Utah.

Given that 60 of the 75 IP-intensive industries were designated as trademark-intensive, it is not surprising to find that 15 of the 16 states with above-average shares of IP-intensive jobs also had above-average shares of trademark-intensive jobs. (See Map 2.)

Only Virginia had a high percentage of IP-intensive jobs, but a below-average share of trademark-intensive employment, as 14.2 percent of employment in the state was in trademark-intensive industries, below the national average of 16.2 percent.

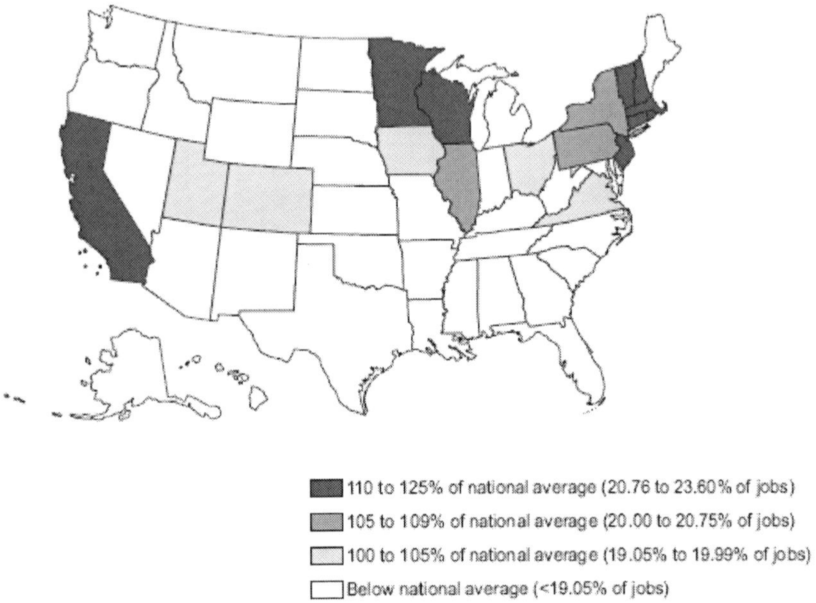

Source: ESA calculations using data from the Bureau of Labor Statistics' Quarterly Census of Employment and Wages.

Map 1. IP-Intensive Industries' Share of Covered Employment by State, 2010.

As highlighted in Map 3, there were 21 states with patent-intensive employment shares above the 3.0 percent national average. States from New England and the Midwest had the highest shares, led by New Hampshire (5.2 percent), Wisconsin (4.9 percent), and Minnesota, Indiana, and Vermont (4.4 percent each).

Essentially all regions of the country had at least one state with an above-average percentage of employment in patent-intensive industries, including the South with North Carolina, South Carolina, and Tennessee.

The 13 states with above-average employment shares in copyright-intensive industries were mostly spread along the East and West coasts, as seen in Map 4.

The District of Columbia (6.4 percent), Virginia (5.9 percent), and New York (4.9 percent) had copyright-intensive employment shares more than 1.5 times above the national average of 3.3 percent.

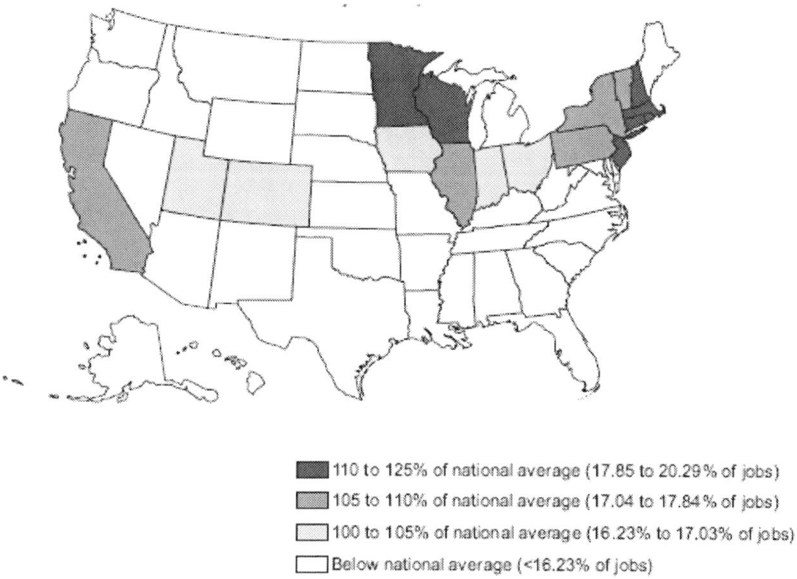

Source: ESA calculations using data from the Bureau of Labor Statistics' Quarterly Census of Employment and Wages.

Map 2. Trademark-Intensive Industries' Share of Covered Employment by State, 2010.

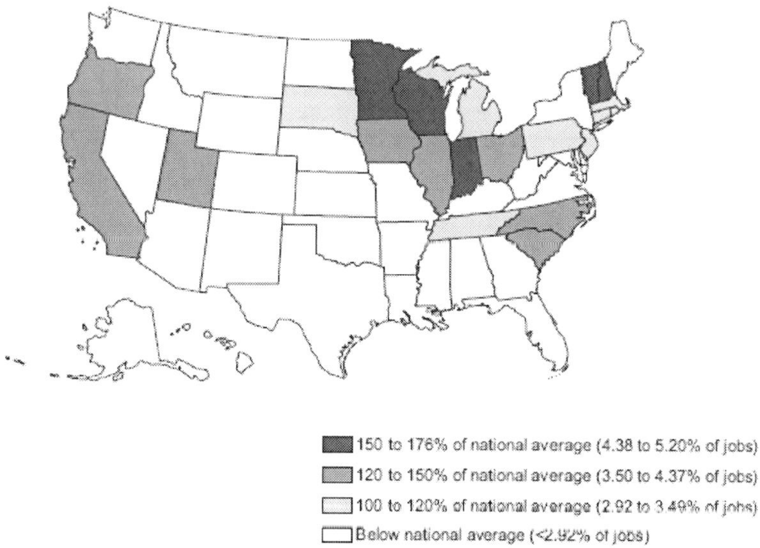

Source: ESA calculations using data from the Bureau of Labor Statistics' Quarterly Census of Employment and Wages.

Map 3. Patent-Intensive Industries' Share of Covered Employment by State, 2010.

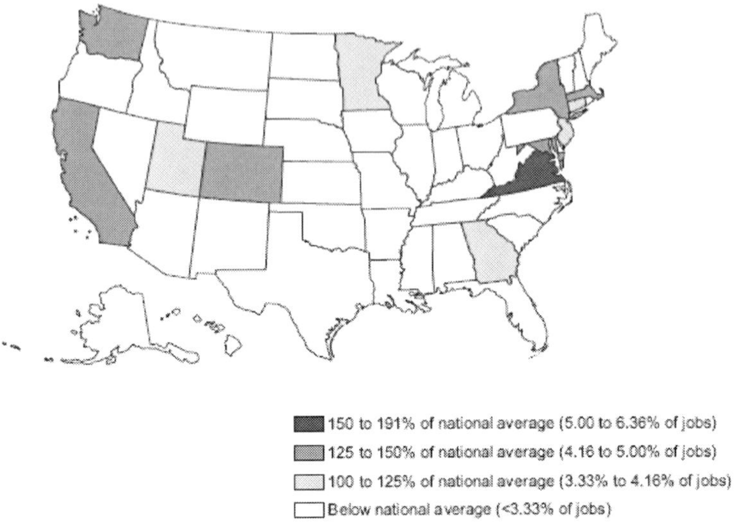

Source: ESA calculations using data from the Bureau of Labor Statistics' Quarterly Census of Employment and Wages.

Map 4. Copyright-Intensive Industries' Share of Covered Employment by State, 2010.

They were followed by Washington State, California, and Colorado (4.7 percent each). Utah and Minnesota were the only other non-coastal states on the list.

Putting these rankings together, we found six states that had above-average employment shares for patent-, trademark-, and copyright-intensive industries: California, Connecticut, Massachusetts, Minnesota, New Jersey, and Utah. These results echo an underlying theme of this report—intellectual property permeates our entire economy. Industries from a broad range of economic sectors are IP-intensive, and the jobs they support span all regions of the United States.

Average Weekly Wages

Private wage and salary workers in IP-intensive industries had average weekly wages of $1,156 in 2010, or 42 percent higher than workers in non-IP-intensive industries in the private sector.[90] (See Figure 6.) This difference was even higher for workers in patent-intensive industries, who earned $1,407 per week on average, and in copyright-intensive industries, with weekly wages of $1,440. Workers in trademark-intensive industries earned $1,111 per week, less than their counterparts in patent-intensive and copyright-intensive industries. This may reflect the fact that trademark-intensive industries represent a wider range of industries than patent-intensive and copyright-intensive industries. As was outlined earlier, patent-intensive industries are a subset of manufacturing industries, while copyright-intensive ones are concentrated in the information sector and the professional and technical services sector. Workers in manufacturing, information, and professional and technical services all have relatively high average weekly wages.

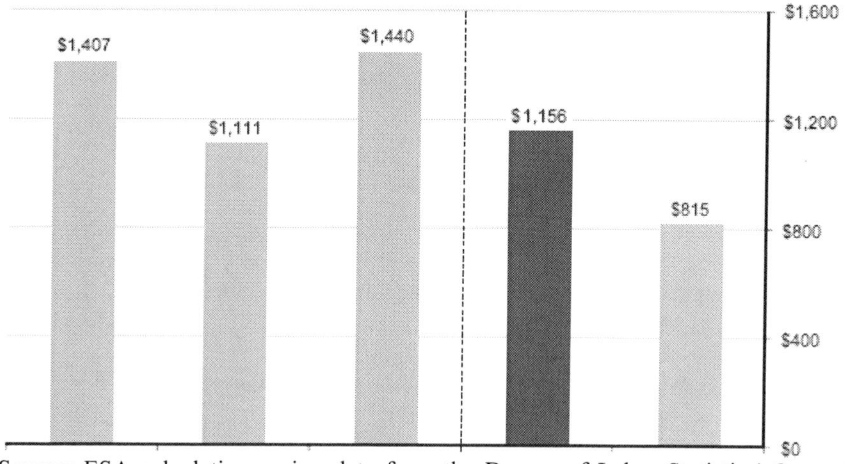

Source: ESA calculations using data from the Bureau of Labor Statistics' Quarterly Census of Employment and Wages.

Figure 6. Average Weekly Wages of Private Wage and Salary Workers in IP-Intensive Industries, 2010.

The premium for working in an IP-intensive industry has grown over time, as shown in Figure 7. In 1990, IP-intensive jobs paid 22 percent more than jobs in other industries and a decade later this premium had risen to 38 percent. It lost some ground early in the 2000s before edging back up and reaching a new high of 42 percent in 2010. Trademark-intensive industries followed a similar but slightly lower upward path over the past two decades. In 1990, trademark-intensive industries paid 20 percent more on average than non-IP-intensive industries, with this premium climbing to 31 percent by 2000 and 36 percent in 2010. Wages in patent-intensive industries started out at a 46 percent premium in 1990, and this premium grew through the 1990s before surging up to 69 percent by 2000. This surge was reversed in 2001, and the premium changed little in the last decade before rising again over the past few years. In 2010, patent-intensive workers earned 73 percent more per hour than workers in non-IP industries.

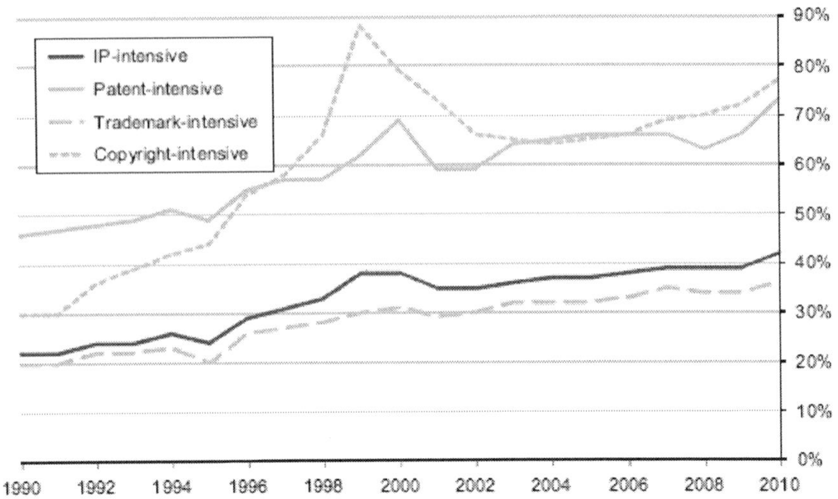

Source: ESA calculations using data from the Bureau of Labor Statistics' Quarterly Census of Employment and Wages.

Figure 7. Average Weekly Wage Premium of Workers in IP-Intensive Industries Relative to Non-IP-Intensive Industries, 1990-2010.

Copyright-intensive industries followed a more extreme version of the trends outlined above. Workers in these industries earned 30 percent more than non-IP workers on average in 1990, and this premium tripled during the following decade to 88 percent in 1999. Over the next five years, the premium decreased to 64 percent before growing again to 77 percent in 2010.

Educational Attainment

In competitive labor markets, wages are closely correlated to worker productivity, and educational attainment is a common gauge of workers' skills and expected productivity. Thus, it would be expected that employees in IP-intensive industries have relatively high educational attainment. The data bear that out as 42.4 percent of workers age 25 and older in the IPintensive industries in 2010 had a bachelor's degree or higher, compared with 34.2 percent in private non-IP-intensive industries.[91] (See Figure 8.) Workers in copyright-intensive industries were the most educated of the three IP-intensive segments with 61.2 percent having attained a bachelor's degree or higher, and only 1.7 percent having less than a high school diploma. Essentially identical shares of workers in the patent- and trademark-intensive industries (38.7 and 38.8 percent, respectively) had at least a bachelor's degree.

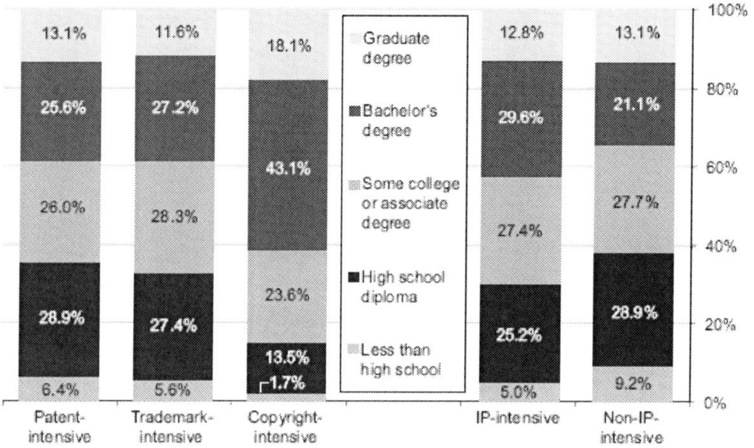

Source: ESA calculations using Current Population Survey public-use microdata.
Note: Estimates are for employed persons age 25 and over.

Figure 8. Distribution of Employed Persons in IP-Intensive Industries by Educational Attainment, 2010.

Foreign Trade

Although trade statistics do not tell us anything about the IP content of exports or imports, U.S. IP-intensive industries are a major part of U.S. trade. Merchandise exports of IP-intensive industries totaled $775 billion in 2010, accounting for 60.7 percent of total U.S. merchandise exports, while merchandise imports of IP-intensive industries stood at $1,336 billion or 69.9 percent of total U.S. merchandise imports. From 2000 to 2010, exports of IP-intensive industries increased 52.6 percent while imports of IP-intensive industries rose 61.6 percent.

Manufacturing industries were responsible for almost 99 percent of IP-intensive merchandise exports in 2010, with oil and gas extraction and software publishing accounting for the rest. Exports from the semiconductors and electronic parts sector led the way, totaling $64.0 billion or 8.3 percent of IP-intensive industries' merchandise exports. (See Figure 9.) The next largest export categories were basic chemicals ($58.4 billion), motor vehicles ($52.4 billion), pharmaceuticals and medicine ($49.4 billion), and computer and peripherals ($44.1 billion).

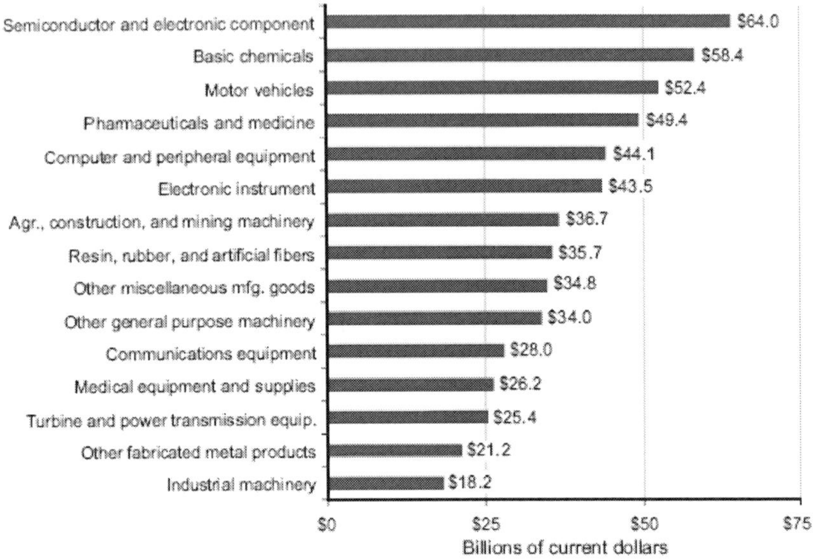

Source: ESA calculations using data from the Census Bureau's Foreign Trade Division.
Note: The selected industries accounted for 74 percent of merchandise exports of IP-intensive industries.

Figure 9. Merchandise Exports of Selected IP-Intensive Industries, 2010.

On the imports side, manufacturing accounted for 79 percent of total imports of IP-intensive industries while oil and gas represented 21 percent and software publishing less than 1 percent. Among imports of manufactured goods, the largest category was motor vehicles (9.7 percent of all imports of IP-intensive industries), followed by computers and peripheral equipment (6.7 percent) and pharmaceuticals and medicines (6.5 percent).

It should be noted that the imports of goods of IP-intensive industries included those produced by U.S.-owned establishments located overseas and by foreign producers under U.S. licenses. In 2009, Americans received $89.8 billion from royalties and license fees, which included industrial processes and trademarks, while paying foreigners $25.2 billion. Imports also provide benefits for American consumers and industries. Imports tend to increase market competition and thus lower prices, making some products affordable to more American consumers.

Likewise, many imports are intermediate inputs for American industries which make their finished products more competitive.[92]

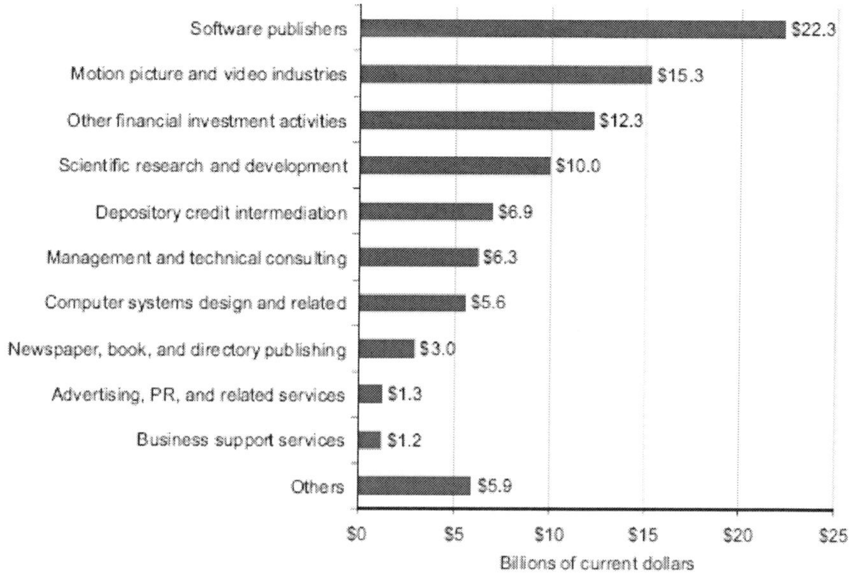

Source: ESA calculations using data from the Census Bureau's 2007 Economic Census.

Figure 10. Exports of IP-Intensive Service-Providing Industries, 2007.

More generally, one must keep in mind that trade statistics do not convey the importance of IP-intensive products to the U.S. economy. For instance, U.S. electronics companies like Apple tend to capture a significant portion of the value added in their global supply chain from products like the iPod and iPhone, two products that are assembled offshore.[93] These companies also tend to capture a large share of the profits from their global supply chains while retaining market share and an innovative edge globally. At home, these companies employ many high-skilled workers who earn much more than the low-wage, low-skill workers that assemble products overseas.[94] Trade statistics themselves do not capture this complete picture of the contribution of these industries to the U.S. economy and its competitiveness. A casual examination of the trade balance in the IP-intensive electronics-related sectors misses this nuanced and important reality.

Data on foreign trade of services are more limited. We used the 2007 Economic Census in order to get the most detailed accounting of services exports and to calculate total exports of IP-intensive service-providing industries. Exports of IP-intensive service-providing industries totaled about $90 billion in 2007, accounting for approximately 19 percent of total U.S. private services exports. As shown in Figure 10, exports of software publishers, at $22.3 billion, were the largest group of services exports in 2007, followed by the motion picture and video industry at $15.3 billion. Other major services export categories in 2007 included financial investment activities ($12.3 billion, excluding securities and commodity contracts intermediation and brokerage), scientific research and development ($10 billion), depository credit intermediation ($6.9 billion), and management and technical consulting ($6.3 billion).

APPENDIX

Revenue Generated from the Licensing of Rights to Use Protected Intellectual Property

Every five years, the Census Bureau conducts an Economic Census of U.S. business establishments, with the latest conducted in 2007. Data collected include number of employees, annual payroll, and value of sales by specific product lines. As a robustness measure of our list of IP-intensive industries, we examined the extent to which industries with high revenue share from IP-related product lines were among our 75 IP-intensive industries. Specifically, we identified 91 product codes from the Economic Census that were associated with licensing, royalties, and other forms of trade of intellectual property.

Examples of such product codes include:

- Licensing of rights:
    - to use intellectual property (product code 39400)
    - to use intellectual property-protected by copyright (39401)
    - to use intellectual property-protected as industrial property (39402)
    - to use intellectual property-protected by trademark (39403)
    - to use intellectual property-protected by patent (39404)
- Outright sale of:
- original works of intellectual property (39250)
- intellectual property protected by copyright (31500)

- "Contract production" of various forms of intellectual property (product codes 30150, 31510, 31520, 35113, 35111, 35114, 35115, 35112, 35110, 35540) and "exclusivity" rights (product code 31256)

Table 11. Industries with IP-Related Revenue, Ranked by IP-Revenue Intensity, 2007

	NAICS code	IP-intensive	Industry title	IP-related revenue ($millions)	Cumulative share	IP-revenue intensity (IP/total revenue)
Above mean; Above mean +1 std. dev.	7115	X	Independent artists, writers, and performers	$9,776	6.8%	76.6%
Above mean; Above mean +1 std. dev.	5331	X	Lessors of nonfinancial intangible assets	24,473	23.7%	74.3%
Above mean; Above mean +1 std. dev.	5121	X	Motion picture and video industries	51,132	59.2%	64.1%
Above mean; Above mean +1 std. dev.	5152	X	Cable and other subscription programming	17,256	71.1%	38.4%
Above mean; Above mean +1 std. dev.	5122	X	Sound recording industries	5,290	74.8%	34.8%
Above mean; Above mean +1 std. dev.	7111	X	Performing arts companies	3,208	77.0%	23.6%
Above mean	5112	X	Software publishers	12,868	86.0%	9.5%
Above mean	5417	X	Scientific research and development services	8,532	91.9%	9.0%
Above mean	5191	X	Other information services	2,812	93.8%	7.5%
Above mean	7113		Promoters of performing arts, sports, and similar events	999	94.5%	6.2%
	7112		Spectator sports	1,024	95.2%	3.4%
	8139		Professional and similar organizations	1,822	96.5%	3.2%
	7114		Agents and managers for public figures	80	96.5%	1.6%
	5511		Management of companies and enterprises	1,349	97.5%	1.3%
	5151	X	Radio and television broadcasting	628	97.9%	1.1%
	6114		Business schools and computer and management training	107	98.0%	1.1%
	5111	X	Newspaper, periodical, book, and directory publishers	883	98.6%	0.6%
	5413		Architectural, engineering, and related services	1,130	99.4%	0.4%
	6117		Educational support services	37	99.4%	0.4%
	8132		Grantmaking and giving services	258	99.6%	0.3%
	8133		Social advocacy organizations	61	99.6%	0.3%
	5419	X	Other professional, scientific, and technical services	86	99.7%	0.2%
	5179	X	Other telecommunications	54	99.7%	0.2%
	8134		Civic and social organizations	23	99.7%	0.2%

Table 11. (Continued)

NAICS code	IP-intensive	Industry title	IP-related revenue ($millions)	Cumulative share	IP-revenue intensity (IP/total revenue)
5415	X	Computer systems design and related services	287	99.9%	0.1%
7121		Museums, historical sites, and similar institutions	20	100.0%	0.1%
6116		Other schools and instruction	19	100.0%	0.1%
5171		Wired telecommunications carriers	19	100.0%	0.0%
5418	X	Advertising, public relations, and related services	16	100.0%	0.0%
5172		Wireless telecommunications carriers (except satellite)	4	100.0%	0.0%
6115		Technical and trade schools	3	100.0%	0.0%
Total		All industries with IP-related revenue	144,259	100.0%	6.6%

Source: ESA calculations using data from the Census Bureau's 2007 Economic Census.

Note: The intensity measure is the percent of overall revenue generated for each four-digit NAICS industry from the licensing of intellectual property protected assets.

Table 11 shows the 31 four-digit NAICS industries that had at least some IP-related revenue, ranked by the IP share of total revenue in 2007. The distribution of IP-related revenue was fairly concentrated. Ten industries had IP revenue shares above the 6.6 percent average (among industries with IP-related revenue), and these industries accounted for about 95 percent of total revenue from IP-related products.

All but one of the ten industries also were identified in this report as IP-intensive.

The sole exception was promoters of performing arts, sports, and similar events (NAICS 7113). This industry generated 6.2 percent of its revenue in 2007 from a few IP-related areas, including exclusivity rights; licensing of rights to use property protected by trademark; licensing of rights to use property protected by copyright; contract production services for intellectual property protected by copyright, excluding live performing arts; and contract production services for intellectual property protected by trademark.[95]

Overall, 14 of the 31 industries with any revenue generated by the protection of IP products and services in 2007 made it into our IP-intensive industry list.

Of the 17 industries that were not defined as IP-intensive, four were in the broader education sector and four were in the "other services (except public administration)" sector.

REFERENCES

Arora, Ashish, Sharon Belenzon, and Luis A. Rios. 2011. "The Organization of R&D in American Corporations: the Determinants and Consequences of Decentralization." NBER Working Paper 17013; *www.nber.org/papers/w17013*

Bureau of Economic Analysis. 1997. *Regional Multipliers: A User Handbook for the Regional Input-Output Modeling System (RIMS II)*; *www.bea.gov/scb/pdf/regional/perinc/meth/rims2.pdf*

Cohen, Wesley M., Richard R. Nelson, and John P. Walsh. 2000. "Protecting Their Intellectual Assets: Appropriability Conditions and Why U.S. Manufacturing Firms Patent (or Not)." NBER Working Paper 7552; *www.nber.org/papers/w7552*

Hall, Bronwyn H., Adam B. Jaffe, and Manuel Trajtenberg. 2001. "The NBER Patent Citation Data File: Lessons, Insights and Methodological Tools." NBER Working Paper 8498; *www.nber.org/papers/w8498*

Hall, Bronwyn H. and Rosemarie H. Ziedonis. 2001. "The Patent Paradox Revisited: An Empirical Study of Patenting in the U.S. Semiconductor Industry, 1979-1995." *RAND Journal of Economics*. 32(1): 101–28.

Higgins, Matthew and Stuart Graham. 2009. "Balancing Innovation and Access; Patent Challenges Tip the Scales." *Science*. Vol. 326.

Interbrand. 2011a. "Best Global Brands;"*www.interbrand.com/en/best-global-brands-methodology/Overview.aspx*

Interbrand. 2011b. "Best Retail Brands;" *www.interbrand.com/en/ Best RetailBrands/2011.aspx*

Jaffe, Adam and Manuel Trajtenberg. 2002. *Patents, Citations and Innovations: A Window on the Knowledge Economy*. MIT Press.

Koopman, Robert, William Powers, Zhi Wang, and Shang-Jin Wei. 2010. "Give Credit Where Credit Is Due: Tracing Value Added in Global Production Chains." NBER Working Paper 16426, *www.nber org/papers/w16426*

Linden, Greg, Jason Dedrick, and Kenneth L. Kraemer. 2009. "Innovation and Job Creation in a Global Economy: The Case of Apple's iPod." Working

Paper, Personal Computing Industry Center, University of California, Irvine; *pcic.merage.uci.edu/papers/2009/InnovationAndJobCreation.pdf*

Linden, Greg, Jason Dedrick, and Kenneth L. Kraemer. 2010. "Who Profits From Innovation in Global Value Chains? A Study of the iPod And Notebook PCs." *Industrial and Corporate Change*. 19(1): 81-116.

Maddison, Angus. 2006. *The World of Economy. A Millennial Perspective (Vol. 1). Historical Statistics (Vol. 2)*. Organization for Economic Co-operation and Development Publications.

Miller, Ronald E. and Peter D. Blair. 1985. *Input-Output Analysis: Foundations and Extensions*. Prentice-Hall, Inc.

National Economic Council, Council of Economic Advisers, and Office of Science and Technology Policy. 2011. *A Strategy for American Innovation: Securing our Economic Growth and Prosperity*; *www.white house.gov/innovation/strategy*

Siwek, Stephen E. 2009. *Copyright Industries in the Economy: the 2003-2007 Report*. Economists Incorporated. Prepared for the International Intellectual Property Alliance (IIPA); *www.ei.com/downloadables/ IIPAS iwekReport2003-07.pdf*

U.S. Copyright Office. U.S. Library of Congress. 2009. *Annual Report of the Register of Copyrights*; *www.copyright.gov/reports/annual/ 2009/ ar2009.pdf*

U.S. Copyright Office. U.S. Library of Congress. 2008. "Copyright Basics;" *www.copyright.gov/circs/circ1.pdf*

U.S. Patent and Trademark Office. U.S. Department of Commerce. 2010a. "Basic Facts about Trademarks;" *www.uspto.gov/trademarks/basics/ BasicFacts_with_correct_links.pdf*

U.S. Patent and Trademark Office. U.S. Department of Commerce. 2010b. *Trademark Manual of Examining Procedure (TMEP)*. 7th edition; *tess2.uspto.gov/tmdb/tmep/*

White House. 2012. *Investing in America: Building an Economy That Lasts*; *www.whitehouse.gov/sites/default/files/investing_in_america_report_final .pdf*

World Intellectual Property Organization. 2003. *Guide on Surveying the Economic Contribution of Copyright-Based Industries*; *www.wipo.int/ copyright/en/publications/pdf/copyright_pub_893.pdf*

End Notes

[1] National Economic Council et al. 2011, 7.

[2] Ibid., 11.

[3] See www.uspto.gov/web/offices/ac/ido/oeip/taf/data/naics_conc/ and also www.google.com/googlebooks/uspto-trademarks.html.

[4] Using data provided by the Bureau of Labor Statistics' Industry Productivity program, employment covers the sum of payroll jobs, self-employed persons, and unpaid family workers, and totaled 144.2 million jobs in 2010. Because the unit of measure is jobs (as opposed to persons) and because about 5 percent of all workers have more than one job, the total number of jobs is greater than the 139.1 million employed persons in 2010, as estimated from the Current Population Survey (www.bls.gov/cps).

[5] Because several industries were found to intensively use both patents and trademarks or copyrights and trademarks, total IP-intensive industry employment is less than the sum of patent-, trademark-, and copyright-intensive industry employment.

[6] Maddison 2006.

[7] National Economic Council et al. 2011, 7.

[8] Ibid., 11.

[9] This analysis uses employment data from the Bureau of Labor Statistics' Industry Productivity program, which includes the sum of payroll jobs, self-employed persons, and unpaid family workers and totaled 144.2 million jobs in 2010. Because the unit of measure is jobs (as opposed to persons) and because about 5 percent of all workers have more than one job, the total number of jobs is greater than the 139.1 million employed persons in 2010, as estimated from the Current Population Survey (www.bls.gov/cps).

[10] In addition to utility patents, there are two other types of patents: design and plant. Utility patents apply to processes, machines, articles of manufacture, composition of matter, or any new and useful improvements thereof. Design patents apply to ornamental designs for an article of manufacture. Plant patents apply to the invention or discovery of selected new varieties of asexually reproducing plants.

[11] 35 U.S.C. ß 157(a)(1). This right was established over 200 years ago in fulfillment of Article 1, Section 8 of the United States Constitution: "The Congress shall have Power... To promote the Progress of Science and useful Arts, by securing for limited Times to Authors and Inventors the exclusive Right to their respective Writings and Discoveries;" www.uspto.gov/patents/index.jsp.

[12] Inventors, who are always individuals and not organizations, apply for patents. The inventors can transfer (assign) their rights to anyone, including organizations, and can do so before or after the patent is issued. Commonly, inventors work for employers who hire them to create new inventions, and as part of their employment agreements agree in advance to assign to their employers any patent rights that result from their work.

[13] Today, multiple inventors on a single patent often come from several nations, an example of globalization's impact on innovation.

[14] The "U.S. corporation" category of ownership in USPTO reports broadly includes private organizations, including small and large companies, nonprofits, partnerships, and universities.

[15] See Tables A1-1a, A1-1b at www.uspto.gov/web/offices/ac/ido/oeip/taf/h_at.htm#PartA1_1b.

[16] Utility patents may be classified into more than one technology class, but are organized according to their primary classification.

[17] This concordance was created by the USPTO with financial support from the National Science Foundation. For an overview of NAICS, see www.census.gov/eos/www/ naics/index.html.

[18] See www.uspto.gov/web/offices/ac/ido/oeip/taf/data/misc/patenting_trends/info_ ptrends 2008.txt.

[19] To contrast the scale of technology patents, design patents break out into 33 classes and plant patents into only one patent class. See www.uspto.gov/web/offices/ ac/ido/oeip/taf/ all_tech.htm for more information on utility patents.

[20] It should be noted that the use of fractional patent counts differs somewhat from our treatment of trademark registrations for which whole counts of registrations by class were used.

[21] As a robustness check, we repeated our analysis using full counts and found little difference in our final results.

[22] For full details on the nature and caveats of the patent data used, see www.uspto.gov/web/ offices/ac/ido/oeip/taf/govt/naics/explan_naics.htm.

[23] Value added and gross output are two alternative gauges of industry size; however, estimates at the level of detail needed for this analysis are not publicly available due to data confidentiality limitations.

[24] In the NAICS classification, a three-digit code is a larger aggregation as compared to a four-digit code. So, for instance, NAICS 236 "construction of buildings" is a larger aggregation of which NAICS 2361 "residential building construction" is a smaller and more specific subset.

[25] Although we pooled the FY 2004-08 patent counts, we also examined individual years to see how stable the selection of patent-intensive industries would be over time. The list of most intensive industries proves to be relatively stable over time, especially the high rankings of the computer and peripheral equipment manufacturers.

[26] See www.uspto.gov/web/offices/ac/ido/oeip/taf/asgstc/usa_ror.htm.

[27] The OneSource Business Browser was used to classify companies to NAICS industries; www.onesource.com/businessbrowser.aspx.

[28] See Cohen et al. 2000.

[29] We note that trademark owners are not required to register their trademarks. Parties can acquire rights in their trademarks merely by using those trademarks. In addition, trademark owners who wish to register their trademark may do so not only at the USPTO, but with state authorities as well, or with both. The trademarks that were analyzed in connection with the present study were all registered at the USPTO.

[30] USPTO 2010a.

[31] For example, the Whole Foods Market grocery store brand name is a service mark for retail grocery store services, and also a trademark for hair shampoo and soap.

[32] There is no further 6-year affidavit requirement after the first 10-year term. Overall, 46 percent of the nearly 822,000 registrations from the 1990s "survived" into the 2000s.

[33] For a list of classes, see Trademark Manual of Examining Procedure (TMEP) chapter 1400 (7th ed. 2010). Fees also are a function of the type of application form used. For more information on the trademarking process, see USPTO 2010a.

[34] An example is the mark "Nike" being registered in classes associated with (a) software, (b) golf equipment, and (c) eyewear. See U.S. trademark registration numbers 3406594, 3389746, and 3081688, each by Nike, Inc.

[35] Another way to think about this is to consider the registration counts to be weighted by the number of classes. So, if the mark for brand A is registered in a single class and the mark for brand B is registered in five classes, then, we calculate total registrations = I (classes per markbrand i * markbrand i) = (1*1) + (5*1) = 6.

[36] This trademark is serial number 76977931 and can be found by using the USPTO searchable trademark database at tarr.uspto.gov/servlet/tarr?regser=serial&entry=76977931.

[37] For example, the list of industries selected because of their trademark intensity, as described in the next section, includes 55 industries. Had we calculated the intensities based on fractional trademark counts, the list would have numbered 51 industries, with 49 industries appearing on both lists.

[38] We could weight companies by number of employees, but it is more difficult to decide whether a given trademark taken by itself should be weighted heavily or lightly.

[39] This methodology has previously been applied in studies of patents, particularly as detailed in Hall, Jaffe, and Trajtenberg 2001. This paper and other supporting documentation are available at www.nber.org/patents/.

[40] To the extent that matched firms differ in their trademarking behavior relative to unmatched firms, these trademark intensities are biased. However, what really matters for our analysis is not whether the estimates are individually biased but rather if any bias varies across industries.

[41] For this reason, this method not only tends to overweight large, publicly traded companies, but also those companies that have a particular trademark-registration approach. Although no studies to our knowledge have been done on the practices of corporate trademark registration at the parent or subsidiary level, recent patent scholarship suggests that IP ownership may show systematic differences along these dimensions. See Arora, Belenzon, and Rios 2011, 29-30.

[42] See www.census.gov/cgi-bin/sssd/naics/naicsrch?code=533110&search=2007, noting that six-digit NAICS industry 533110 and four-digit NAICS industry 5331 are identical. The entire 2007 NAICS manual can be found online at www.census.gov/eos/www/naics/index.html.

[43] See www.uspto.gov/about/stratplan/ar/index.jsp to access the annual "Performance and Accountability Reports." In each report, the Top 50 Trademark Registrants tables are Table 29B within the USPTO Workload Tables in section 5, "Other Accompanying Information."

[44] This subscription-based dataset is available online at businessbrowser. onesource. com/homepage.aspx.

[45] As the results will highlight, classifying a company into a single NAICS industry is not straightforward. Generally, this report used the OneSource entry for the parent company of the firm in the top 50 listing to assign an industry code. The exceptions were when a company did not appear in OneSource and when the listed firm's industry differed considerably from the parent company's industry. As an example of the latter case, Bath & Body Works is a health and personal care store (NAICS 4461) even though its parent company Limited Brands, Inc. is classified under clothing stores (NAICS 4481). Other publicly available data were used to classify companies not included in OneSource.

[46] Toys, jewelry, sporting goods, pencils, signs, musical instruments, buttons, and caskets, among other goods, are some of the product lines in the miscellaneous manufacturing industry. Games and toys, however, are what dominate the trademarking activity for this NAICS code, as Hasbro, Mattel, International Game Technology, and WMS Gaming are each classified within "Other miscellaneous manufacturing." These four companies account for 18 of the 32 appearances for this NAICS code in the top 50 lists.

[47] Three companies were responsible for the bulk of the top 50 trademark registrations in the motion picture and video industry: Twentieth Century Fox, Warner Brothers, and World Wrestling Entertainment. The fact that these three companies are classified principally in the motion picture and video industry helps illustrate the difficulty of summarizing large enterprises' activities within a single industry code.

[48] See USPTO Bulk Data available at www.google.com/googlebooks/uspto.html.

[49] This analysis was restricted to U.S.-owned firms because of the great difficulty in assigning NAICS codes to small, foreign-owned firms.

[50] Because the sample was drawn by registrations as opposed to registrants, a single individual or company could be drawn more than once. One company was selected twice in the sample, and so the 196 registrations corresponded to 195 firms.

[51] Corporate brand generally is equivalent to company name (with its corresponding protected logo), as opposed to a specific product brand, with some exceptions, such as Sprite and Budweiser.

[52] Interbrand 2011a.

[53] Wal-Mart does top the separate rankings of the best retail brands in 2011. See Interbrand 2011b.

[54] The difficulty of assigning a single NAICS code to companies is further highlighted in our analysis of these companies. Take, for example, General Electric, which as described in OneSource is a "diversified technology, media and financial services company... [with] products and services includ[ing] aircraft engines, power generation, water processing, security technology, medical imaging, business and consumer financing, media content and industrial product." No one NAICS code can capture GE. In terms of sales, GE Capital is largest of the company's five principal segments, accounting for 32 percent of total revenue in calendar year 2010, and by this criteria, it makes sense to classify GE in nondepository credit intermediation (NAICS 5222). Still, this does miss the fact that the vast majority of GE's revenues is from other sectors. Two of those sectors, interestingly, also cover industries selected as trademark-intensive. The NBC Universal business segment operates in radio and television broadcasting (NAICS 5151) while part of the Home & Business Solutions segment includes household appliance manufacturing (NAICS 3352). Together these two segments account for 17 percent of total revenue. It should be noted that this discussion applies to data prior to GE's sale of NBC Universal to Comcast.

[55] For a complete listing of Interbrand's top 100 Global Brands, see Interbrand 2011a.

[56] Although Budweiser is part of Anheuser-Busch InBev N.V., which is based in Leuven, Belgium, Interbrand considers the Budweiser brand to be American.

[57] This trademark is serial number 85255019 and can be found by using the online search engine at tarr.uspto.gov/servlet/tarr?regser=serial&entry=85255019.

[58] U.S. Copyright Office 2008, 2.

[59] A work is "created" when it is fixed in a copy for the first time, regardless of whether the work is published or not.

[60] U.S. Copyright Office 2008, 7.

[61] See the Brief History of the Copyright Office at www.copyright.gov/circs/circ1a.html. Copyright registration data are published in U.S. Copyright Office 2009, 68.

[62] Ibid., 69.

[63] World Intellectual Property Organization 2003.

[64] See, for example, Siwek 2009.

[65] World Intellectual Property Organization 2003, 29. The core copyright industries represent one of four main groups of copyright-based industries. The others are interdependent, partial, and non-dedicated support industries.

[66] This discussion should not imply that distribution industries as a whole cannot by our definitions be considered IPintensive. As discussed above, a broad range of industries seek trademark protection, including distribution industries like clothing stores, which are identified as IP-intensive.

[67] World Intellectual Property Organization 2003, 27.

[68] These data are from the Bureau of Labor Statistics, Occupational Employment Statistics program and are for private wage and salary workers in newspaper, periodical, book and directory publishers (NAICS 5111). See www.bls.gov/oes/current/naics4_511100.htm#27-0000.

[69] Jaffe and Trajtenberg 2002.

[70] The pharmaceutical industry is a prime example of the former. Merck & Company, for instance, lost billions of dollars in sales beginning in 2008 on the osteoporosis drug Fosamax when generic manufacturers successfully invalidated its core patent. See Higgins and Graham 2009, 370-71.

[71] Electronics products are prime examples of the latter, as products from Motorola, Samsung, and Nokia, now competing with Apple's iPhone in the hand-held market, all require the rights to many patents in order to be legally sold in the U.S. market.

[72] The appendix to this report examines the extent to which industries selected as IP-intensive have relatively high revenue shares from the licensing of rights to use protected intellectual property and other IP-related product lines.

[73] Cohen et al. 2000.

[74] Hall and Ziedonis 2001.

[75] Unpaid family workers also are included in the employment estimates from the BLS Industry Productivity program. However, because unpaid family workers account for only about one percent of the combined total of the self-employed and unpaid family workers outside of agriculture, they are not referred to or analyzed elsewhere in this report.

[76] As noted earlier, the employment data cited here were provided by the Bureau of Labor Statistics' Industry Productivity program. Employment covers the sum of payroll jobs, self-employed persons, and unpaid family workers and totaled 144.2 million jobs in 2010. Because the unit of measure is jobs (as opposed to persons) and because about 5 percent of all workers have more than one job, the total number of jobs is greater than the 139.1 million employed persons in 2010, as estimated from the Current Population Survey (www.bls.gov/cps).

[77] The 4.5 million jobs figure can be calculated by subtracting overall IP-intensive employment from the sum of employment of the trademark-, patent-, and copyright-intensive industries.

[78] Non-IP-intensive employment refers to employment in private sector industries not selected as IP-intensive, plus employment in government and in private households.

[79] Data on industrial capacity published by the Federal Reserve Board and data on private fixed investment by industry from the Bureau of Economic Analysis' Fixed Assets Accounts Table 3.8ES, as cited in White House 2012, 1 and 7. The data are available at www.federal reserve.gov/releases/g17/caputl.htm and www.bea.gov/iTable/index_FA.cfm.

[80] Data from the Bureau of Labor Statistics' Current Employment Statistics program, available at www.bls.gov/ces.

[81] The copyright-intensive industries from the information sector are newspaper, periodical, book and directory publishers; software publishers; motion picture and video industries; sound recording industries; radio and television broadcasting; cable and other subscription programming; and other information services. The copyright intensive industries drawn from the professional and technical services sector are specialized design services; computer systems design and related services; advertising, public relations, and related services; and other professional, scientific, and technical services.

[82] See Bureau of Economic Analysis' National Income and Product Accounts, Table 1.1.6, available online at http://www.bea.gov/iTable/index_nipa.cfm.

[83] Data on industrial production are published by the Federal Reserve Board, with historical data available at www.federalreserve.gov/releases/g17/table1_2.htm.

[84] Data from the Bureau of Labor Statistics' Current Employment Statistics program, available at www.bls.gov/ces.

[85] These estimates are derived from unpublished input/output tables computed and analyzed by staff from the Bureau of Economic Analysis' Industry Accounts office. Data are consistent with the December 2010 annual revision to the annual input/output tables and are based on the 2009 after-redefinition domestic make and use tables and estimates of the share of 2010 employment attributable to final demand in IP-intensive industries. Total output requirement tables were calculated based on the make and use table following the methodology published in mathIO.doc, which are available at www.bea.gov/ industry /zip/cxctr2002detail.zip.

[86] In Miller and Blair 1985, these multipliers are referred to as "type I employment multipliers," while in the Bureau of Economic Analysis 1997, they are referred to as direct-effect employment multipliers.

[87] The 16.2 million jobs in IP-intensive industries directly linked to final demand was found by examining the share of IP output associated with final demand (excluding imports). We then fed the vector of final demand IP-intensive employment into a jobs-to-jobs total requirements matrix in several iterations until we accounted for all 27.1 million jobs in the IP-intensive industries. The first iteration told us that a total of 29.4 million jobs were associated with the 16.2 million final-demand, IP-intensive jobs, giving us the 1.8 multiplier figure we cite. This first iteration, however, only accounted for about 20 million jobs in IP-intensive industries. Subsequent iterations uncovered the full set of supply chain jobs in IP-intensive industries as well as the full supply chain in other (non-IP-intensive) industries.

[88] Value added is defined as the difference between an industry's total output (its sales plus the change in inventories arising from production) and the value of its intermediate purchases from other industries (that is, from its supply chain). GDP can be defined as the sum of value added across all industries in the economy. When value added is summed across all industries, industry sales to and purchases from each other cancel out, and the remainder is industry sales to final users, or GDP.

[89] The data used to examine state employment are from the Bureau of Labor Statistics' Quarterly Census of Employment and Wages (QCEW) and measure covered employment. Covered employment refers to jobs covered by state and Federal unemployment insurance law, and includes practically all civilian wage and salary employment. These data do not include the self-employed or unpaid family workers. For more information on QCEW, see www.bls.gov/cew.

[90] This section focuses on the average weekly earnings of private wage and salary workers using data from the QCEW. According to QCEW data, there were 24.5 million private wage and salary jobs in IP-intensive industries in 2010. Wages represent total compensation paid during the calendar year, regardless of when services were performed. Included in wages are pay for vacation and other paid leave, bonuses, stock options, tips, the cash value of meals and lodging, and in some states, contributions to deferred compensation plans (such as 401(k) plans).

[91] These estimates are calculated from 2010 Current Population Survey public-use microdata, as accessed through the Census Bureau's DataFerrett tool at dataferrett.census.gov.

[92] An estimated 53 percent of U.S. merchandise imports were intermediate products in 2007, according to unpublished data from Koopman et al. 2010.

[93] Linden, Dedrick, and Kraemer 2010.

[94] Linden, Dedrick, and Kraemer 2009.
[95] See factfinder.census.gov/servlet/IBQTable?_bm=y&-ds_name=EC0771I3&-NAICS2007=7113&-_lang=en.

In: Intellectual Property
Editor: Nicolas Vallée

ISBN: 978-1-62417-313-4

Chapter 2

INTELLECTUAL PROPERTY RIGHTS AND INTERNATIONAL TRADE*

Shayerah Ilias and Ian F. Fergusson

SUMMARY

This chapter provides background on intellectual property rights (IPR) and discusses the role of U.S. international trade policy in enhancing IPR protection and enforcement abroad. IPR are legal rights granted by governments to encourage innovation and creative output by ensuring that creators reap the benefits of their inventions or works, and they may take the form of patents, trade secrets, copyrights, trademarks, or geographical indications. U.S. industries that rely on IPR contribute significantly to U.S. economic growth, employment, and trade with other countries. Counterfeiting and piracy in other countries may result in the loss of billions of dollars of revenue for U.S. firms as well as the loss of U.S. jobs. Responsibility for developing IPR policy, engaging in IPR-related international negotiations, and enforcing IPR laws cuts across several different U.S. government agencies.

Promoting the enforcement of IPR is an important component of U.S. international trade policy. Since the 1995 Agreement on Trade-Related Aspects of Intellectual Property Rights (TRIPS Agreement) at the World Trade Organization (WTO), trade policy has been used to promote

* This is an edited, reformatted and augmented version of a Congressional Research Service publication, CRS Report for Congress RL34292, dated February 17, 2011.

enforcement of IPR abroad. The United States and several trading partners have been negotiating the Anti-Counterfeiting Trade Agreement (ACTA), which would surpass TRIPS Agreement commitments.

The United States also pursues international IPR support through regional and bilateral free trade agreements (FTAs), which often include IPR commitments by U.S. partners exceeding their TRIPS Agreement obligations. However, the May 10, 2007 bipartisan trade agreement led to a scale-back of some of the IPR requirements in the Peru, Panama and Colombia FTAs, in an effort to bolster bipartisan support for the FTAs. Other trade policy tools also are available for U.S. efforts to advance international IPR. Pursuant to Section 182 of the Trade Act of 1974 as amended (P.L. 93-618), the Office of the U.S. Trade Representative (USTR) identifies countries providing inadequate IPR protection in its annual "Special 301" report. Section 337 of the amended Tariff Act of 1930 authorizes the U.S. International Trade Commission (ITC) to prohibit U.S. imports of infringing products. Additionally, under the Generalized System of Preferences (GSP), the United States may consider a developing country's IPR policies and practices as a basis for offering or suspending preferential duty-free entry to certain products from the country.

IPR protection and enforcement has been a focal point of legislative activity in recent sessions of Congress. In the 110th Congress, legislation was enacted to establish a new entity to coordinate intellectual property activities within the federal government (P.L. 110-403). In the 111th Congress, legislation was introduced calling for greater U.S. international IPR enforcement efforts and increased prioritization of resources devoted to such activities (H.R. 496 and related bill S. 1466; H.R. 2410 and H.R. 2475).

Given the role of IPR in the U.S. economy and its contribution to U.S. employment and trade, IPR issues related to international trade policy may figure prominently in the 112th congressional agenda. Congress may choose to consider whether or not FTAs are an appropriate vehicle for boosting intellectual property protection and enforcement. Congress also may balance IPR protection and enforcement with other public policy goals such as access to medicine in poor or developing countries. In addition, Congress may examine the effectiveness of the current U.S. coordinating structure for promoting international IPR support.

INTRODUCTION

Intellectual property rights (IPR) traditionally have been matters of national concern. Individual nation states have developed IPR regimes reflecting their domestic needs and priorities. Over time, intellectual property protection and enforcement have come to the forefront as a key international trade issue for the United States, figuring prominently in the multilateral trade policy arena and in regional and bilateral U.S. free trade agreements (FTAs).

The protection and enforcement of IPR in the United States and abroad is of key interest to Congress. Intellectual property is an increasingly critical component of the U.S. economy. Industries that rely on intellectual property protection in the United States claim to lose billions of dollars each year due to overseas IPR infringement. In light of the recent international financial crisis and global economic downturn, congressional interest has grown in the role of IPR in advancing U.S. industrial competitiveness and contributing to U.S. economic recovery following the recent international crisis. Members of Congress also have expressed concern about the potential health and safety consequences of counterfeit pharmaceutical drugs and other products, as well as the possible link between terrorist groups and traffic in counterfeit and pirated goods.

This chapter discusses the different kinds of IPR and forms of IPR infringement; importance of IPR to the U.S. economy; estimated losses associated with IPR infringement; organizational structure of IPR protection in multilateral, regional, bilateral arenas; U.S. government agencies involved with IPR and trade; and issues for Congress regarding IPR and international trade.

Intellectual Property Rights Basics

This section provides definitions of the various kinds of intellectual property rights (patents, trade secrets, copyrights, trademarks, and geographical indications) and intellectual property rights misappropriation (infringement, piracy, and counterfeiting).

Types of IPR

IPR are legal rights granted by governments to encourage innovation and creative output. They ensure that creators reap the benefits of their inventions or works and may take the form of patents, trade secrets, copyrights, trademarks, or geographical indications. Through IPR, governments grant a temporary legal monopoly to innovators by giving them the right to limit or control the use of their creations by others. IPR may be traded or licensed to others, usually in return for fees and or royalty payments. Although the World Trade Organization (WTO) Agreement on Trade-Related Aspects of Intellectual Property Rights (TRIPS) provides minimum standards for IPR protections, such rights are granted on a national basis and are, in general, enforceable only in the country in which they are granted. However, countries are obliged to abide by WTO rules and their IPR enforcement practices can be challenged by other countries at the WTO.

Patents

The Patent Act (35 U.S.C. 101 *et seq*) governs the issuance and use of patents in the United States. Patents are granted for inventions of new products, processes, or organisms (known as utility patents). Patents also may be granted for designs and plants. For an invention to be patentable, it must be new, "non-obvious" (involving an inventive step), and have a potential industrial or commercial application. The patent provides the holder with the exclusive right to sell the invention for a period of 20 years, or to prevent the incorporation of the invention into other products without the permission of the rights-holder. The patent right is based on the proposition that inventors must be granted a temporary monopoly over their invention in order to encourage innovation and to promote the expenditure of money on research and development. The patent holder recoups these up-front costs through a temporary monopoly over sale of the invention. In return for this economic rent, the patent holder must disclose the content of the patent along with test data and other information concerning the invention. This is meant to spur further creativity by those seeking to build on the patent after its expiration. Domestically, patents are granted by the Patent and Trademark Office (PTO) of the Department of Commerce.

Trade Secrets

A trade secret is any type of valuable information, including a "formula, pattern, compilation, program device, method, technique, or process," that derives independent economic value from not being generally known or readily ascertainable and is subject to reasonable efforts by the owner to maintain its secrecy.[1] Examples of trade secrets include blueprints, customer lists, and pricing information. While protection of patents and copyright is a matter of federal law, trade secret protection is found also in state law. However, most states subscribe to the Uniform Trade Secret Act (UTSA).

There are important differences between trade secrets and patents. Individuals do not have to apply for trade secret protection as they would for patents. Protection of trade secrets originates immediately with the creation of the trade secret; there is no process for applying for or registering trade secrets. Trade secret protection does not expire unless the trade secret becomes known. In contrast, patent applicants must disclose information about their innovation to the PTO in order to acquire a patent. Patents offer rights holders stronger protection but for a limited period of time. While applying for a patent can be a costly and lengthy process, patents are valuable if the confidentiality of the innovation is fragile or if the area of research is highly competitive.

Copyright

Protection of copyrights in the United States is based on the Copyright Act (17 U.S.C. 101, *et seq*). Copyrights protect original expressions of authorship. Such protections include literary or artistic works such as books, music, sound recordings, movies, paintings, architectural works, and computer software and databases (though not individual bits of data). Traditionally, copyrights differed from patents in that there was no claim to industrial applicability or novelty of the idea. The expression of the idea, not the underlying idea, was being copyrighted. While some of the criteria for copyrights differ from those of patents, the objective is the same: investments of time, money, and effort to create work of cultural, social and economic significance should be protected to encourage further creativity. U.S. law protects authorship for life plus 70 years for personal works, or 120 years from creation (or 95 years from publication) for corporate works. Copyrights may be registered by the Copyright Office of the Library of Congress, or acquired through creating and fixation of the work of authorship.

Trademarks

Trademark protection in the United States is governed jointly by state and federal law. The main federal statute is the Lanham Act of 1946 (15 U.S.C. 1051, *et seq*). Also known as service marks, trademarks permit the seller to use a distinctive name, mark, or symbol to identify and market a product, service, or company. The trademark allows quick identification of the seller's product, and for good or ill, can become an indicator of a product's quality. If for good, the trademark can be valuable in the introduction of new products by conveying an instant assurance of quality. The trademark is designed to prevent other companies with similar merchandise from free-riding on the association of quality with the trademarked item. Thus, a trademarked good may command a premium in the marketplace because of its reputation. For trademarks, distinctiveness is at a premium because a trademark must capture the consumer's imagination to be effective as generic names of commodities cannot be trademarked. Trademark rights are acquired through use or through registration with the PTO.

A related concept to trademarks is the geographic indication, which is also protected by the Lanham Act. The geographic indication acts to protect the quality and reputation of a distinctive product originating in a certain region; however, the benefit does not accrue to a sole producer, but rather the producers of a region. Geographic indications are generally sought for agricultural products, or wines and spirits. Protection for geographical indications is acquired in the United States by registration with the PTO, through a process similar to trademark registration.[2]

Infringement of IPR

IPR infringement is the misappropriation or violation of the IPR. In the case of patents, infringement of a patent owner's exclusive rights (as afforded by patent laws) involves a third party's unauthorized use of the patented device. As relates to international trade, the greatest challenge to the patent right is infringement in foreign countries, or non-observance by WTO member states to the minimal standards of the TRIPS Agreement. Copyright infringement occurs when a third party engages in reproducing, performing, making sound or visual recordings of, or broadcasting a copyrighted work without the consent of the copyright owner.

Piracy

The term "piracy" has applications to both copyrights and trademarks. The major challenge facing copyright protection is piracy, either through physical duplication of the work, illegal dissemination of copyrighted material (such as computer software, music, or movies) over the Internet, and/or participation in commercial transactions of copyrighted materials without the consent of the copyright owner. With respect to trademarks, piracy involves the registration or use of a famous foreign trademark that is not registered in the country or is invalid because the trademark has not been used.

Counterfeiting

An imitation of a product is referred to as a "counterfeit" or a "fake." Counterfeit products are manufactured, marketed, and distributed with the appearance of being the genuine good and originating from the genuine manufacturer.[3] The purpose of counterfeit goods is to deceive consumers about their origin and nature. Counterfeiting and copying of original goods are major challenges for trademarked products. The counterfeited product can be sold for a premium because of its association with the original item, while reducing the sales of the original items. Furthermore, consumer experience with a counterfeited good of inferior quality, can damage the reputation of the trademark product. Popular examples of counterfeit products in fake fashionwear, such as Louis Vuitton bags or Rolex watches, or fake pharmaceutical products, such as popular brand-name prescription medicines.

A related issue is the imitation of labels and packaging of trademarked goods. In this situation, the imitator uses a trademark that is confusingly similar to a well-known trademark in order to benefit from the reputation of the product with which he is competing.

GLOBAL INTELLECTUAL PROPERTY HOLDINGS

The total number of patent filing applications received under the Patent Cooperation Treaty (PCT), an international patent filing system administered by the World Intellectual Property Organization (WIPO), has grown in recent years. After peaking in 2008, international patent filings under the PCT fell by 4.5% in 2009, reflecting the international financial crisis. The contraction of the global economy has been associated with a decline in investment and spending on research and development.[4] In 2010, international patent filings grew by 4.8% to 162,900—nearly 2008 levels (see Table 1).[5]

Intellectual property holdings that are protected by international agreements are highly concentrated in certain countries. The United States continues to be the source of the world's largest number of patent filing applications under the PCT, accounting for nearly one-third of such filings in 2010. However, the U.S. growth rate of patent filings has been negative in recent years. The United States, along with Germany and Japan, accounted for about 60% of all patent applications filed in 2010 under the PCT. China ranked as the fourth largest source of international patent filings under the PCT in 2010, representing about 8% of global filings. China had the highest growth rate in such filings, at about 56% in that year.

Table 1. Global Intellectual Property Filings Through the PCT, 2006-2009.

Country	2008	2009		2010	
	Filings	Filings	Filings	Percent of	Growth from
Country	2008	2009		2010	
				Total	Previous Years
United States	51,637	45,618	44,855	27.5%	-1.7%
Japan	28,760	29,802	32,156	19.7%	7.9%
Germany	18,855	16,797	17,171	10.5%	2.2%
China	6,120	7,900	12,337	7.6%	56.2%
South Korea	7,899	8,035	9,686	5.9%	20.5%
France	7,072	7,237	7,193	4.4%	-0.6%
United Kingdom	5,466	5,044	4,857	3.0%	-3.7%
Netherlands	4,363	4,462	4,097	2.5%	-8.2%
Switzerland	3,799	3,671	3,611	2.2%	-1.6%
Sweden	4,137	3,567	3,152	1.9%	-11.6%
Canada	2,976	2,527	2,707	1.7%	7.1%
Italy	2,883	2,652	2,632	1.6%	-0.8%
Finland	2,214	2,123	2,076	1.3%	-2.2%
Australia	1,938	1,740	1,736	1.1%	-0.2%
Spain	1,390	1,564	1,725	1.1%	10.3%
All Others	13,725	12,659	12,909	7.9%	2.0%
Total	**163,234**	**155,398**	**162,900**		

Source: World Intellectual Property Organization.

CONTRIBUTION OF INTELLECTUAL PROPERTY TO U.S. ECONOMY

Intellectual property is an important source of comparative advantage for the United States. Nearly every industry depends on IPR for its businesses. Among the industries that are dependent on patent protection are the aerospace, automotive, computer, consumer electronics, pharmaceutical, and semiconductor industries. Copyright-based industries include the software, data processing, motion picture, publishing, and recording industries. Other industries that indirectly benefit from IPR protection include retailers, traders, and transportation businesses, which support the distribution of goods and services derived from intellectual property.[6]

The role of IPR in the U.S. economy has been longstanding. Some evidence suggests that factors linked to innovation account for about three-fourths of the United States' post-World War II growth rate.[7] In recent years, the role of IPR in the U.S. economy has grown. Various studies suggest that IP-related industries are one of the largest source of jobs in the United States. One study using data from 2000-2007 found that, among tradable industries, IP-intensive industries surpass non-IP-intensive industries on a range of economic measures, including job creation, wages, output and sales per employee, and exports. During this time period, according to the study, IP-intensive industries paid both their low- and highly skilled employees close to 60% more than non-IP-intensive industries. The report also found that IP-intensive industries represented nearly 60% of total U.S. exports during 2000-2007.[8] More broadly, IPR-intensive industries may contribute positively to the U.S. economy through productivity gains and other spillover effects.

Industry-specific figures may further demonstrate the importance of IP to the U.S. economy. For example, in 2007, the business and entertainment software, motion picture, recording, and publishing industries, which rely on copyright protection, were estimated to contribute about $889 billion to the U.S. economy ("value-added" to current GDP), or about 6.44% of the U.S. economy. This was an increase from 2006, during which the value-added of these copyright industries to the U.S. GDP totaled $837 billion, or 6.35% of the U.S. economy. These copyright industries also accounted for nearly 23% of real U.S. annual economic growth in 2007, up from about 13% in 2006. In terms of U.S. employment, the copyright industries represented 4% of U.S. workers (5.6 million workers) in 2007, similar to the prior year. Foreign sales and exports from these industries amounted to $126 billion in 2007, up from $116 billion in 2006.[9]

The pharmaceutical industry, which is patent-intensive, provides another illustration of intellectual property contributions to the U.S. economy. In 2009, domestic sales by research-based pharmaceutical companies that are members of Pharmaceutical Researchers and Manufacturers of America (PhRMA) reached an estimated $183 billion, while sales abroad by PhRMA member companies totaled about $103 billion.[10]

The intellectual property industries contribute positively to the overall U.S. trade balance through royalties and licensing fees. Rights-holders may authorize the use of technologies, trademarks, and entertainment products that they own to entities in foreign countries, resulting in revenues through royalties and license fees.[11] In 2009, U.S. receipts from cross-border trade in royalties and license fees (relating to patent, trademark, copyright, and other intangible rights) totaled $89.8 billion, down from $93.9 billion in the previous year. Also in 2009, U.S. payments of royalties and license fees to foreign countries amounted to $25.2 billion, down from $25.8 billion in the year before. Industrial processes, computer software, and trademarks accounted for the bulk of U.S. international trade in intangible assets.[12]

Some advocates of civil liberties assert that empirical analysis on the role of IPR in the U.S. economy may not be fully evaluating the economic and commercial benefits of lawful exceptions and limitations to exclusive rights. For example, by one estimate, businesses that rely on "fair use" exceptions to U.S. copyright law contribute $2.2 trillion to the U.S. economy. The "fair use" doctrine permits limitations and exceptions to the exclusive right afforded by copyright law. It permits limited use of copyrighted works without requiring permission from the right holder in certain cases, examples of which may include news reporting, research, teaching, and library use.[13]

Prevalence and Economic Consequences of IPR Infringement

Advances in information and technology and declining costs of transportation and communication, spurred by globalization, have fundamentally changed information and trade flows. Such changes have created new markets for U.S. exporters, but at the same time, have been associated with the proliferation of counterfeiting and piracy on a global scale.

Several factors contribute to the growing problem of IPR infringement. While the costs and time for research and development are high, IPR infringement is associated with relatively low costs and risks and a high profit margin. According to PhRMA, it takes a pharmaceutical company about 10 to 15 years of research and development to create a new drug. PhRMA member companies collectively spent an estimated $46 billion for research and development (domestic and abroad) in 2009.[14] In contrast, drug counterfeiters can lower production costs by using inexpensive, and perhaps dangerous or ineffective, ingredient substitutes.

The development of technologies and products that can be easily duplicated, such as recorded or digital media, also has led to an increase in counterfeiting and piracy. Increasing Internet usage has contributed to the distribution of counterfeit and pirated products. Additionally, civil and criminal penalties often are not sufficient deterrents for piracy and counterfeiting. The United States is especially concerned with *foreign* IPR infringement of U.S. intellectual property. Compared to foreign countries, IPR infringements levels in the United States are estimated to be relatively low.

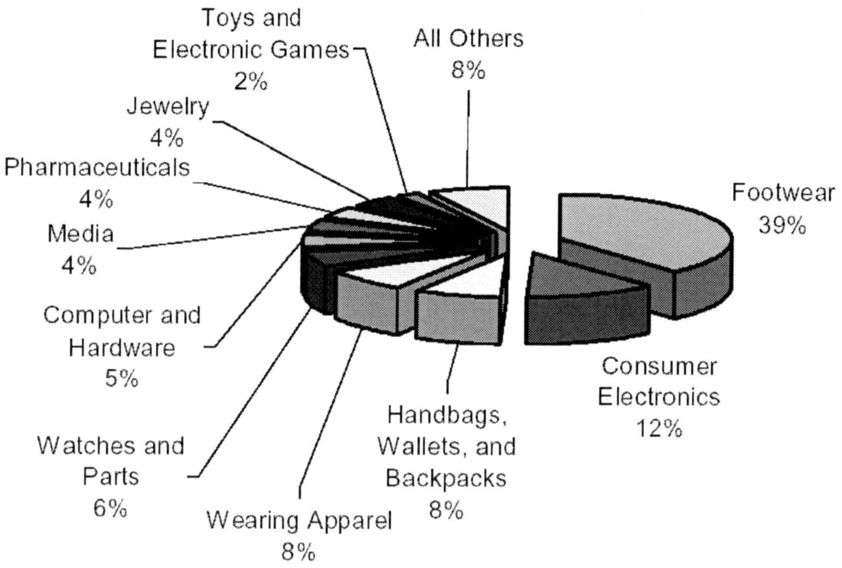

Sources: U.S. Customs and Border Protection and U.S. Immigration and Customs Enforcement.

Figure 1. Border Seizures of Counterfeit and Pirated Goods: FY2009.

Seizures

Because of the secretive, illicit nature of IPR infringement, it is difficult to estimate the magnitude of its impact on U.S. producers and exporters. However, customs data on seizures of counterfeit and pirated goods may offer some idea of the magnitudes involved. One study by the Organization for Economic Cooperation and Development (OECD) indirectly extrapolated available customs data on seizures to conclude that world trade in counterfeit and pirated goods may have amounted to about $200 billion in 2005. In particular, the study used the customs information to estimate the probability that imports of particular goods from particular countries would be pirated or counterfeit. The OECD estimate does not include the counterfeit and pirated goods produced and consumed within a country and does not include infringing goods distributed over the Internet. If these figures were included, the trade estimate likely would be higher.[15] Updated estimates from the OECD suggest that trade in IPR-infringing goods may have totaled up to $250 billion in 2007. During that same time period, the share of counterfeiting and pirated goods in world trade also is estimated to have increased—from 1.85% in 2000 to 1.95% in 2007.[16]

Data on pirated and counterfeit seizures of imports at the U.S. border shed light of the magnitude of the issue in the U.S. context (see **Figure 1**). In FY2009, the Customs and Border Protection (CBP) and Immigration and Customs Enforcement (ICE) agencies of the Department of Homeland Security (DHS) made 14,841 IPR-related seizures, more than double the FY2005 level of 8,022. Between FY2005-FY2008, the domestic value of IPR-related seizures grew by more than 25% each year. The domestic value of seizures peaked at $272 million in FY2008 and then dropped by 4% to about $261 million in FY2009.[17]

Of all U.S. trading partners, China continues to account for the majority of counterfeits intercepted at the U.S. border. In FY2009, seizures of goods originating from China represented 79% of all seizures and $205 million in value. Other top trading partners from which IPR-infringing goods were seized include Hong Kong, India, Taiwan, and Korea. [18]

A top priority for the CBP is seizing counterfeit imports that endanger the health and safety of consumers, such as fake healthcare products, pharmaceutical products, and consumer electronics. The total value of IPR-related seizures of commodities that represent potential safety and security risks was $32 million in FY2009. Pharmaceutical goods were the top commodity posing safe and security risks, accounting for 34% ($11 million in domestic value) of such commodities intercepted at U.S. borders. Imports from China represented 62% of IPR-related goods that were intercepted at U.S. borders and identified as presenting safety and security risks, followed by such imports from India at 9%.[19]

Sectoral Infringement

U.S. industries that rely on IPR protection claim to lose billions of dollars in revenue annually due to piracy and counterfeiting. In addition, beyond the direct losses faced by U.S. intellectual property-based firms, the U.S. economy may face additional "downstream" losses from IPR infringement. According to this view, counterfeiting and piracy losses to U.S. firms, for example, also result in the loss of jobs that would have been created if the infringement did not occur, which translates into lost earnings by U.S. workers. This, in turn, translates into lost tax revenues for federal, state, and local governments from lost personal income, corporate income, and production taxes.[20]

Attempts have been made in some economic sectors to quantify the IPR infringement levels and related losses to legitimate U.S. businesses. Two intellectual property-based sectors that have calculated the extent of infringement in their industries and related costs are copyrights and pharmaceuticals. A discussion of estimated losses from these two sectors follows.

Copyright Industry

The International Intellectual Property Alliance (IIPA), a coalition of seven member associations representing over 1,900 U.S. copyright-based companies, provides annual estimates of U.S. trade loss associated with copyright infringements in selected countries.[21] For 2009, IIPA estimated that copyright piracy in 43 countries resulted in $14.3 billion in U.S. business software losses and $1.4 billion in U.S. records and music losses. China was the leading culprit in terms of trade losses due to copyright piracy, contributing to $3.4 billion in business software losses and $466 million in records and music losses to U.S. businesses.[22]

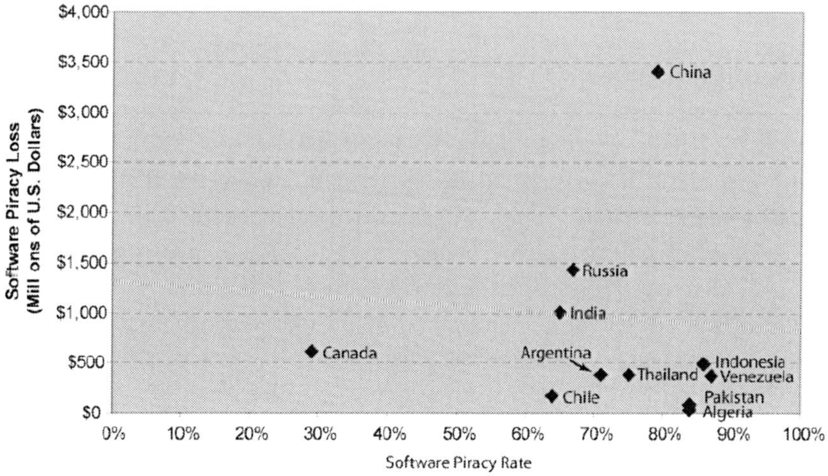

Source: Business Software Alliance.

Notes: The countries listed were identified as Priority Watch List countries in the 2010 "Special 301" report released by the Office of the U.S. Trade Representative.

Figure 2. Estimated Software Piracy Rates in Selected Countries and Estimated Related Software Piracy Losses to U.S. Businesses.

There is not always a direct relationship between IPR infringement rates and the costs to U.S. firms (see **Figure 2**). The size of a country's market and U.S. industries' access to the market affect the extent to which infringement rates translate into costs to U.S. IPR-based industries. Thus, estimated losses from IPR infringement may be lower in a country with high piracy rates but a small market for the United States, compared to a country with lower piracy rates but a bigger market for United States.

Table 2. Estimated Damages for PhRMA Member Companies From Data Exclusivity and Patent Protection Violations.

	2006 Submission (Oct. 2004 - Sept. 2005)	**2007 Submission (Oct. 2005 - Sept. 2006)**
Total Damages	\$13.9 million	\$21.7 million
Total Sales	\$74.6 million	\$172.1 million
Damages as a Percentage of Sales	19%	13%

Source: PhRMA 2006 and 2007 Special 301 Submissions.

Pharmaceutical Industry

In previous years, PhRMA has provided annual estimates of U.S. pharmaceutical industry losses from foreign violations of data exclusivity and patent protection.[23] In its 2007 Special 301 submission to the USTR, PhRMA contended that its member companies sustained damages totaling an estimated \$21.7 million from data exclusivity and patent violations in 24 countries (see Table 2).[24] Damages reported in the 2007 submission were nearly double those reported in the prior year's submission. While the total loss associated with IPR infringement grew, damages as a percentage of sales declined from the 2006 submission to the 2007 submission.[25] At the time of reporting for submissions for subsequent years, PhRMA was not able to provide estimates of damages resulting from trade barriers associated with intellectual property protection and market access.[26]

Limitations on Data Estimating IPR Infringement Costs

There may be limitations to data estimating the impact of counterfeiting and piracy on the U.S. economy. According to a study conducted by the U.S.

Government Accountability Office (GAO), the U.S. government does not systematically collect data or analyze the impacts of counterfeiting and piracy on the U.S. economy. In many cases, the federal government relies on estimates conducted by industry groups. However, industry associations may not always release their proprietary data sources and methods, complicating efforts to verify such estimates.[27] Some critics point out that many of the estimates for losses associated with IPR infringement are generated by industry groups that may have self-interested motivations.[28]

There is a possibility that such IPR infringement loss estimates may overestimate the extent to which sales of pirated and counterfeit goods displace legitimate sales. Methods for calculating data on counterfeiting and piracy often involve certain assumptions. Estimates of losses from IPR infringement can be highly sensitive to how these assumptions are derived and weighted. Some analysts question the proposition that sales of pirated goods translate directly into revenue losses for legitimate firms. The basic economic model employed in such estimates assumes that there is perfect substitutability between pirated and legitimate goods, which would equate sales of pirated goods to revenue losses of legitimate U.S. copyright businesses. Some analysts suggest that legitimate firms face a competition threat *only* if the individuals purchasing counterfeit products would be able and willing to purchase the legitimate product at the price offered when piracy is not present.[29] For consumers in poor developing countries, especially, this assumption may not be tenable. Others also question the statistical techniques used to develop these estimates.[30]

At the same time, there is a possibility that some IPR infringement loss estimates may underestimate the losses to U.S. businesses. For example, U.S. trade losses due to copyright infringement may be higher than reported because estimates oftentimes do not account for all forms of piracy, such as Internet piracy, which is an increasingly significant contributor to copyright piracy.[31] One study estimates that nearly 24% of global internet traffic infringes upon copyright.[32]

THE ORGANIZATION STRUCTURE OF IPR PROTECTION

Given the importance of intellectual property to the U.S. economy and the economic losses associated with counterfeiting and piracy, the United States is a leading advocate of strong global IPR standards and enforcement. Increasingly, the United States has integrated IPR policy in its international trade policy activities, pursuing enhanced IPR laws and enforcement through multilateral, regional and bilateral trade agreements, and national trade laws.

Multilateral IPR System

World Trade Organization (WTO)

At the center of the present multilateral trading system is the World Trade Organization (WTO), an international organization established in 1995 as the successor to the General Agreements on Tariffs and Trade (GATT). The WTO was established as the result of the Uruguay Round of trade negotiations (1986-1994), which resulted in numerous agreements on trade in goods, services, investment and other non-tariff barriers to trade. One of the Uruguay Round agreements was the Agreement on Trade-Related Aspects of Intellectual Property Rights (TRIPS Agreement), which sets minimum standards on intellectual property rights protection and enforcement with which all WTO member states must comply. The United States, the European countries, and the IPR business community were instrumental in including IPR on the Uruguay Round agenda. Many developing countries were wary of including IPR in trade negotiations, preferring to discuss them under the World Intellectual Property Organization (WIPO) (see below) instead. However, developing countries acceded, after being granted delayed compliance periods, and after achieving negotiating goals on other issues such as textiles and clothing, and savoring the prospect of operating under a rules-based trading system.

While previous international agreements on intellectual property rights continue to exist (see Table 2), the TRIPS Agreement was the first time that intellectual property rules were incorporated into the multilateral trading system. Two basic tenets of the TRIPS Agreement are national treatment (signatories must treat parties of other WTO members no less favorably in terms of IPR protection than the party's own nationals) and most-favored-nation treatment (any advantage in IPR protection granted to the party of another WTO member shall be granted to nationals of all other WTO member states).

Much of the TRIPS Agreement sets out the extent of the agreement's coverage of the various types of intellectual property: copyrights, trademarks, geographical indications, industrial designs, patents, layout of circuitry design, trade secrets, and test data. The TRIPS Agreement provisions build on several existing IPR treaties administered by the WIPO (discussed below). Another part provides standards of enforcement for IPR covered by the agreement. It enumerates standards for civil and administrative procedures and remedies, the application of border measures, and criminal procedures. A Council for the TRIPS Agreement was established to monitor the implementation of the agreement and transition arrangements were devised for developing countries. Finally, the agreement provides for the resolution of disputes under the Uruguay Round Agreement's Dispute Settlement Understanding (DSU). The binding nature of the DSU, with the possibility of the withdrawal of trade concessions (usually the re-imposition of tariffs) for non-compliance, sets this agreement apart from previous IPR treaties that did not have effective dispute settlement mechanisms.

U.S. WTO Cases against China on IPR

In April 2007, the United States filed two WTO dispute settlement cases against China, alleging inadequacies in China's enforcement of IPR laws and its barriers to market access for U.S. copyright businesses.[33]

- In January 2009, the DSU issued its final ruling on the case centering on IPR enforcement issues. The WTO panel ruled in the United States' favor that China's denial of copyright protection to works without censorship approval is inconsistent with the TRIPS Agreement. The panel also agreed with the United States that it is impermissible for China to publicly auction IPR-infringing goods seized at the border, with the only requirement being that fake brands and trademarks be removed from the goods. The WTO panel ruled that more evidence was needed before deciding whether or not China's threshold values for prosecuting counterfeiting and piracy permit commercial scale IPR infringement. China agreed to implement the WTO ruling.[34]
- In August 2009, a WTO panel ruled that a number of China's restrictions on trading rights and distribution of IPR-related products were inconsistent with WTO rules. However, the WTO panel did not address whether China's censorship policies or import limitations on foreign films violate WTO rules. China agreed to implement the WTO ruling.[35]

The TRIPS Agreement also seeks a balance of rights and obligations between the private right, enumerated above, and the obligation "to secure social and cultural development that benefits all."[36] Article 7 declares that:

> ... the protection and enforcement of IPR should contribute to the promotion of technological innovation and to the transfer and dissemination of technology, to the mutual advantage of producers and users of technological knowledge and in a manner conducive to social and economic welfare and to a balance of rights and obligations.

This paragraph attempts to link the protection of IPR with greater technology transfer, including technology covered by IPR protection, to the developing world. The language itself has been interpreted in various ways. Developed countries have tended to consider this language exhortatory, but developing countries have tried, without much success, to make technology transfer a meaningful obligation within the TRIPS Agreement system. Article 66.2 of the agreement requires developed country members to provide incentives to their enterprises and institutions to promote technology transfer to least-developed countries to assist them in establishing a viable technology base. Developed countries report annually on their efforts to encourage technology transfer (LDCs).

Complying with international IPR standards may impose greater burdens on developing countries than developed countries. Developing countries generally have to engage in greater efforts to bring their laws, judicial processes, and enforcement mechanism into compliance with the TRIPS Agreement. Consequently, developing countries were given an extended period of time in which to bring their laws and enforcement mechanisms into compliance with the TRIPS Agreement. Developing countries and post-Soviet states were given an additional four years from the entry into force of the agreement (January 1, 1995). For products that were not covered by a country's patent system (such as pharmaceuticals in many cases), an additional five years was granted to bring such products under coverage. For developing countries, all provisions of the TRIPS agreement should now be in force. For the least developed countries (LDCs), the phase-in period was set at 10 years (January 1, 2006), and for pharmaceuticals, the compliance period was later extended to 2016.[37]

Declaration on TRIPS Agreement and Public Health [38]

In agreeing to launch the Doha Round of WTO trade negotiations, trade ministers adopted a "Declaration on the TRIPS Agreement and Public Health" on November 14, 2001.[39] The Declaration sought to alleviate developing country dissatisfaction with aspects of the TRIPS regime. It delayed the implementation of patent system provisions for pharmaceutical products for least developed countries (LDCs) until 2016. The declaration committed member states to interpret and implement the agreement to support public health and to promote access to medicines for all. The Declaration recognized certain "flexibilities" in the TRIPS agreement to allow each member to grant compulsory licenses for pharmaceuticals and to determine what constitutes a national emergency, expressly including public health emergencies such as HIV/AIDS, malaria, and tuberculosis or other epidemics.

Paragraph 6 of the Declaration directed the WTO members to formulate a solution to a corollary concern, the use of compulsory licensing by countries with insufficient or inadequate manufacturing capability. Compulsory licenses are issued by governments to authorize the use or production of a patented item by a domestic party other than a patent holder. They are authorized by Article 31 of TRIPS, which places certain limitations on their use, scope, duration. A provision that predominantly restricted production authorized by compulsory license to the domestic market became the focal point of the negotiations because it, in effect, conveys the right of compulsory licensing only to countries with the capability to manufacture a given product. Countries without a domestic manufacturing capability were essentially precluded from using this flexibility of the TRIPS agreement.

On the eve of the Cancun Ministerial in August 2003, WTO members agreed on a Decision[40] to waive the domestic market provision of the TRIPS article on compulsory licensing (Article 31(f)) for exports of pharmaceutical products for "HIV/AIDS, malaria, tuberculosis and other epidemics" to least developed countries (LDCs) and countries with insufficient manufacturing capacity. This Decision was incorporated as an amendment to the TRIPS agreement at the Hong Kong Ministerial in December 2005.

The amendment must be ratified by two-thirds of the 153 WTO member states. The deadline for ratification has been extended to December 30, 2011. Until then, the 2003 waiver continues in force. To date, the following WTO members have ratified the amendment: the United States, Switzerland, El Salvador, South Korea, Norway, India, the Philippines, Israel, Japan, Australia, Singapore, Hong Kong, China, the 27 countries of the European Union, Mauritius, Egypt, Mexico, Jordan, Brazil, Morocco, Albania, Macau, Canada, Bahrain, Colombia, Zambia, Nicaragua, Pakistan, the Former Yugoslav Republic of Macedonia, Uganda, Mongolia, Croatia, and Senegal.[41]

The system established by the WTO allows LDC and countries without sufficient manufacturing capacity to issue a compulsory license to a company in a country that can produce such a product. After a matching compulsory license is issued by the producer country, the drug can be manufacturing and exported subject to various notification requirements, quantity and safeguard restrictions. While several exporting countries have established laws and procedures for implementing this system, only Rwanda has availed itself to use the system to import HIV/AIDS medicines from a generic manufacturer in Canada.[42]

Intellectual Property Protection and Development

The controversy over the relationship between IPR and development was engaged by the advent of the TRIPS Agreement, which for the first time placed IPR obligations on developing countries. Some hold that expansion of IPR is an obstacle to growth and development in less advanced countries, while others, with a diametrically opposing view, maintain that IPR are beneficial to both developed and developing countries.

Some IPR critics believe that a strong IPR regime may reduce developing countries' access to technology from advanced countries by imposing higher fees for technology licenses and production right, limiting their innovation and economic growth and development.

For instance, Japan, Singapore, Taiwan, and South Korea enhanced their technological abilities and developed their economies through "reverse engineering" of foreign technologies.

Others claim that IPR promote technology transfer through increased trade, foreign investment, and licensing in the long-run by making a country more attractive to foreign partners. A 2002 OECD study concluded that stronger IPR laws, particularly enhanced patent standards, may be associated with increased foreign direct investment (FDI) and trade for developing countries over time, with variation by industries and level of development.[43] For instance, India experienced an increase in foreign investment and technology transfer once it expanded its patent protection. China offers a counterexample of a country with a weak IPR regime but high FDI and trade levels.

There is also evidence that IPR's impact on developing countries may vary by the level of development. One study suggests that IPR protection may offer more benefits for the more industrialized developing countries, such as Brazil and India, compared to other developing countries. Such industrializing economies could experience economic growth of as much as 0.5% annually through increased trade, FDI, and licensing.[44]

Another study finds that rapid economic growth is associated with weak intellectual property regimes, but that developing countries with higher levels of per capita income may benefit economically from stronger IPR regimes.[45]

There is also concern that strengthened patent protection may drive up prices for medicines or delay the entry of generic drugs into the market, reducing access to HIV/AIDS treatments and other drugs. IPR supporters argue that strong IPR is critical to creating incentives for pharmaceutical innovations and suggest that reduced prices are no guarantee that needed goods will make it into the hands of individuals in developing countries due to political corruption, poverty, and poor social infrastructure.

World Intellectual Property Organization (WIPO)

In addition to the WTO, the other main multilateral venue for addressing IPR issues is the World Intellectual Property Organization (WIPO), a United Nations agency. Established in 1967, WIPO is charged with fostering the effective use and protection of intellectual property globally. WIPO's mandate focuses exclusively on intellectual property, in contrast to the WTO's broader international trade mandate.

WIPO's antecedents are the 1883 Paris Convention for the Protection of Industry Property and the 1886 Berne Convention for the Protection of Literary and Artistic Work. Most of the substantive provisions of these two treaties are incorporated in the WTO's TRIPS Agreement. WIPO's primary function is to administer a group of IPR treaties which put forth minimum standards for member states (shown in Table 3). All international IPR treaties, save TRIPS, are administered by WIPO.

In order to address digital technology issues not dealt with in the TRIPS Agreement, WIPO established the WIPO Copyright Treaty (WCT) and WIPO Performance and Phonograms Treaty (WPPT) in 1996.[46] Recent WIPO efforts have focused on patent law.

In June 2000, WIPO signatories adopted the Patent Law Treaty (PLT), which called for harmonization of patent procedures. This agreement went into force on April 28, 2005. Discussions began in May 2001 for the Substantive Patent Law Treaty (SPLT), which targets issues specifically related to patent grants, but stalled in 2006. Government leaders participating in the Group of 8 (G8) meeting in July 2008 called for "accelerated discussions" of the SPLT.[47]

Table 3. Summary of WIPO-Administered IPR Treaties

Treaty	Date Concluded	Provisions
Intellectual Property Protection Treaties		
Paris Convention for the Protection of Industrial Property (Paris Convention)	1883 (entered into force 1884)	Protects industrial property (includes patents, marks, industrial designs, utility models, trade names, and geographic indications)
Berne Convention for the Protec-tion of Literary and Artistic Works (Berne Convention)	1886 (entered into force 1886)	Protects literary and artistic works, providing right to control and receive payments for use
Madrid Agreement for the Repre-ssion of False and Deceptive Indications of Source on Goods (Madrid Agreement - Indications of Source)	1891	Requires States to seize imported goods with false/deceptive indications of source or to prohibit importation of such goods; open to States party to Paris Convention (1883)
Rome Convention for the Protec-tion of Performers, Producers of Phonograms and Broadcasting Organizations (Rome Convention)	1961	Protects rights of performers against certain acts to which they have not agreed; protects rights of producers of phonograms, and broadcasting organizations to authorize/prohibit certain acts; open to States party to Berne Convention (1886)
Convention for the Protection of Producers of Phonograms Against Unauthorized Duplication of their Phonograms (Phonograms Convention)	1971	Protects producers of phonograms against unauthorized reproduction of their phonograms or importation of duplications for public distribution
Brussels Convention Relating to the Distribution of Programme-Carrying Signals Transmitted by Satellite (Brussels Convention)	1974	Protects against the unauthorized distribution of program-carrying signals transmitted by satellite
Nairobi Treaty on the Protection of the Olympic Symbol (Nairobi Treaty)	1981	Protects Olympic symbol against unauthorized commercial uses
Treaty on the International Registration of Audiovisual Works (Film Register Treaty)	1989	Establishes International Register for Audiovisual Works

Treaty	Date Concluded	Provisions
Treaty on Intellectual Property in Respect to Integrated Circuits (Washington Treaty)	1989	Protects layout designs which display electrical components of an integrated circuit
Trademark Law Treaty (TLT)	1994	Streamlines national and regional trademark registration processes
WIPO Copyright Treaty (WCT)	1996 (entered into force 2002)	Special agreement under Berne Convention; grants exclusive rights to owners of copyright in computer programs and compilations of data/other material
WIPO Performances and Phonograms Treaty (WPPT)	1996 (entered into force 2002)	Grants exclusive rights to performers and phonogram producers
Patent Law Treaty (PLT)	2000 (entered into 2005)	Aims to harmonize and streamline national and regional patent applica-tion procedures and patents
Singapore Treaty on the Law of the Trademarks	2006 (not yet in force)	Builds on TLT (1994); aims to harmo-nize trademark registration procedures; has wider scope (includes communi-cation technology developments)
Global Protection System Treaties		
Budapest Treaty on the Interna-tional Recognition of the Deposit of Microorganisms for the Purposes of Patent Procedure (Budapest Treaty)	1977 (entered into force 1980)	Special agreement under Paris Convention (1883); requires States to recognize the deposit of a microorg-anism with any "international depositary authority"
Madrid Agreement Concerning the International Registration of Marks (Madrid Agreement -Marks)	1891	Requires seizure of imported goods with false/deceptive indication of source or prohibition of importation of such goods; open to States party to Paris Convention (1883)
Hague Agreement Concerning the International Registration of Industrial Designs (Hague Agreement)	1925 (entered into force 1928)	Allows protection of industrial designs in all member states on basis of single application with WIPO; three acts currently in force: 1934, 1960, and 1999 Acts
Lisbon Agreement for the Protection of Appellations of Origin and their International Registration (Lisbon Agreement)	1958	Provides international protection for geographical indications

Table 3. (Continued)

Treaty	Date Concluded	Provisions
Patent Cooperation Treaty (PCT)	1970 (entered into force 1978)	Establishes an international patent filing system; allows a single international patent application to have legal standing in all countries signatory to PCT; open to States party to Paris Convention (1883)
Protocol Relating to the Madrid Agreement (Madrid Protocol)	1989 (entered into force 1995)	Relates to Madrid Agreement (1891); seeks to make Madrid system more amenable to domestic laws of certain who are not yet signatories to Madrid Agreement; open to States party to Paris Convention (1883)
Classification Treaties		
Nice Agreement Concerning the International Classification of Goods and Services of the	1957 (entered into force 1961)	Establishes a classification of goods and services in order to register trademarks and service marks; open to
Purposes of the Registration of Marks (Nice Agreement)		States party to Paris Convention (1883)
Locarno Agreement Establishing an International Classification for Industrial Designs	1968 (entered into force 1971)	Establishes a classification for industrial designs; open to States party to Paris Convention (1883)
Strasbourg Agreement Concerning the Industrial Patent Classification (Strasbourg Agreement)	1971 (entered into force 1975)	Establishes the International Patent Classification (IPC); open to States party to Paris Convention (1883)
Vienna Agreement Establishing Classification of the Figurative Elements of Marks (Vienna Agreement)	1973 (entered into force 1985)	Establishes a classification for marks which consist/contain figurative components; open to States party to Paris Convention (1883)

Source: WIPO.

Table 4. Patent and Copyright Provisions in the TRIPS Agreement and U.S. FTAs

Intellectual Property Forms	TRIPS Provisions	General TRIPS-Plus Provisions in FTAs	Scale-down of TRIPS-Plus Standards
Patents			
Patent term extensions	No provisions	Mandatory extensions in cases of unreasonable delays in patent grants/ regulatory approval *Jordan (Article 4.23.a), Chile (Article 17.9.6; 17.9.2a), Singapore (Article 16.7.7; 18.8.4a), Australia (Article 17.9.8;17.10.4), Morocco (Article 15.9.7; 15.10.3), CAFTA-DR (Article 15.9.6; 15.10.2), Bahrain (Article 14.8.6),Oman (Article 15.8.6), Korea (Article 18.8.6)*	Optional extensions in cases of unreasonable delays in patent grants/Regulatory approval *NAFTA (Article 1709.12)* *Peru (Article 16.9.6), Panama (Article 16.9.6), Colombia (Article 16.9.6)*
Market approval linked to patent status	No provisions *NAFTA (no mention), Jordan (no linkage, but patent owner must be notified if another entity is seeking marketing approval for generic version of patented product, Article 4.23.b)*	National regulatory authorities cannot provide marketing approval for a generic version of a patented drug without permission from rights-holder; also requires notification of rights-holder if marketing permitted *Chile (Article 17.10.2b), Singapore (16.8.4c), Australia (Article 17.10.4), Morocco (Article 15.10.4), CAFTA-DR (Article 15.10.2), Bahrain (Article 14.9.4), Oman (15.9.4), Korea (Article 18.9.5)*	Eliminates mandate that regulatory authorities cannot approve a generic drug for marketing if patent for drug in place *Peru (Article 16.10.4), Panama (Article 15.10.4), Colombia (Article 16.10.4)*

Table 4. (Continued)

Intellectual Property Forms	TRIPS Provisions	General TRIPS-Plus Provisions in FTAs	Scale-down of TRIPS-Plus Standards
Protection for undisclosed test or other data	Members must protect data from unfair commercial use (Article 39.3) *Jordan (Article 4.22)*	Provides for at least five years of data exclusivity from date of approval in country for pharmaceuticals that contain new chemical products *NAFTA (Article 1711.6), Bahrain (Article 14.9.1), Oman (Article 15.9(1-2), CAFTA-DR (Article 15.10.1), Singapore (Article 16.8(1-3)), Australia (Article 17.10.1), Morocco (Article 15.10.1), Chile (Article 17.10.1), Korea (Article 18.9(1-2))*	Provides for at least five years of marketing exclusivity from date of approval in country of first filing if new drug is granted marketing approval within six months in country of second filing *Peru (Article 16.10.2), Panama (Article 15.10.4), Colombia (Article 16.10.2)*
Issuance of compulsory licenses	Some restrictions in issuance of compul-sory licenses; circumstances under which licenses can be issued not limited (Article 13) *NAFTA (Article 1709.10),*	Limits issuance of compulsory license to specific cases: Correcting anti-competitive practices, public noncommercial contexts, national emergencies, and other extremely urgent situations *Jordan (Article 4.20), Singapore (Article 16.7.6), Australia (Article 17.9.7)*	Not discussed *Chile (no mention), Morocco (no mention), CAFTA-DR (no mention), Bahrain (no mention), Oman (no mention) Peru (no mention), Panama (no mention), Colombia (no mention), (no mention)*

Intellectual Property Forms	TRIPS Provisions	General TRIPS-Plus Provisions in FTAs	Scale-down of TRIPS-Plus Standards
Parallel importing of patented products	TRIPS will not be used to discuss IPR exhaustion (Article 6) *Jordan (no mention), Chile (no mention), CAFTA-DR (no mention), Bahrain (no mention), Oman (no mention)*	Parallel importation can be restricted or prohibited *NAFTA (Article 1709.5, 1709.9), Singapore (Article 16.7.2), Morocco (Article 15.9.4), Australia (Article 17.9.4)*	Not discussed *Peru (no mention), Panama (no mention), Colombia (no mention), Korea (no mention)*
Biodiversity and traditional knowledge	Members may exclude plants and animals from patentability (microorganisms and non-biological and micro-biological processes must be eligible for patents); must provide protection of plant varieties (Article 27.3(b))	Countries shall make patents available for plants and animals *Morocco (Article 15.9.2, plants 7and animals mentioned, plant varieties are not mentioned)*	Members may exclude plants and animals from patentab-ility, but shall take reasonable effort to provide patent protection for plants or animals and maintain protection once offered *Chile (Article 17.9.2, mentions plants but not animals), CAFTADR (Article 15.9.2), Peru (Article 16.9.2), Panama (Article 15.9.2), Colombia (Article 16.9.2)*
	NAFTA (Article 1709.3), Bahrain (Article 14.8.(1-2)), Oman (Article 15.8.2, plants not discussed), Jordan, (no mention), Singapore (no mention), Australia (no mention), Korea (no mention)		

Table 4. (Continued)

Intellectual Property Forms	TRIPS Provisions	General TRIPS-Plus Provisions in FTAs	Scale-down of TRIPS-Plus Standards
Copyrights			
Rights-management information	Not discussed *NAFTA (no mention), Jordan (no mention)*	Outlaws removal or alternation of information *Chile (Article 17.5.6), Australia (Article 17.4.8), Singapore (Article 16.4.8), Morocco (Article 15.5.9), CAFTA-DR (Article 15.5.8), Bahrain (Article 14.4.8), Oman (Article 15.4.8), Peru (Article 16.7.5), Panama (Article 15.5.8), Colombia (Article 16.7.5), Korea (Article 18.4.8)*	
Term of protection	No less than 50 years from authorized publication (Article 12) *NAFTA (Article 1705.4), Jordan (no mention)*	No less than 70 years from death of author or authorized publication *Chile (Article 17.5.4), Singapore (Article 16.4.4), Australia (Article 17.4.4), Morocco (Article 15.5.5), CAFTA-DR (Article 15.5.4), Bahrain (Article 14.4.4), Oman (Article 15.4.4), Peru (Article 16.5.5), Panama (Article 15.5.4), Colombia (Article 16.5.5), Korea (Article 18.4.4)*	
Circumvention of copyrighted work	Not discussed *NAFTA (no mention)*	Signatories must agree to prohibit circumvention *Jordan (Article 4.6), Chile (Article 17.5.5), Singapore (Article 16.4.7), Australia (Article 17.4.7), Morocco (Article 15.5.8), CAFTA-DR (Article 15.5.7), Bahrain (Article 14.4.7), Oman (Article 15.4.7), Peru (Article 16.7.4), Panama (Article 15.5.7), Colombia (Article 16.7.4), Korea (Article 18.4.7)*	
ISP Liability	Not discussed *NAFTA (no mention), Jordan (no mention)*	ISPs are provided with limited liability in certain situations of copyright infringement on their servers if they comply with regulations *Chile (Article 17.11.23), Singapore (Article 16.9.22), Australia (Article 17.11.29), Morocco, CAFTA-DR (Article 15.11.27), Bahrain, Oman (Article 15.10.29), Peru (Article 16.11.29), Panama (Article 15.11.27), Colombia (Article 16.11.29), Korea (Article 18.10.30)*	

Note: When there is no mention of an issue in an FTA, the TRIPS standard generally holds.

WIPO's other functions include assisting member states through training programs, legislative information, intellectual property institutional development, automation and office modernization efforts, and public awareness activities. WIPO's enforcement activities are more limited than those of the WTO. Through its Advisory Committee on Enforcement (ACE), WIPO cooperates with member states to promote international coordination on enforcement activities.

With the emergence of the TRIPS Agreement, some observers question the relevance of WIPO. However, others contend that the TRIPS Agreement has given WIPO a new and stronger role. Through a 1996 agreement between the WTO and WIPO, the two organizations have agreed to work closely together to ensure the implementation of the TRIPS Agreement by member states through legal and technical assistance and technical cooperation.[48] In 1998, WIPO and WTO began a joint initiative based on the 1996 agreement to enhance their coordination of technical cooperation activities in order to assist developing countries, in particular, to fulfill their TRIPS commitments.[49]

Free Trade Agreements

In recent years, the United States increasingly has focused on free trade agreements (FTAs) as an instrument to promote stronger IPR regimes by foreign trading partners. In general, the United States has viewed the TRIPS Agreement and WIPO-administered treaties as a minimum standard and has pursued higher IPR protection and enforcement levels through regional and bilateral agreements.

Trade Promotion Authority and Negotiating Objectives

Under Trade Promotion Authority (TPA), Congress delegates its constitutional authority to regulate foreign commerce to the President to negotiate and enter into certain free trade agreements (FTAs), and to have their implementing bills considered under expedited legislative procedures (no amendment, up-or-down vote), provided the President follows the guidelines, objectives, reporting, and consultation requirements mandated by Congress. IPR have become important negotiating objectives in grants of trade promotion authority; the most recent extension of that authority expired on July 1, 2007.

IPR negotiating objectives for FTAs were first enacted in trade promotion authority (then known as fast-track authority) by the Omnibus Trade and Competitiveness Act of 1988 (P.L. 100-418).

The act sought enactment and enforcement of adequate IPR protection from negotiating partners. It also sought to strengthen international rules, dispute settlement, and enforcement procedures through the GATT and other existing intellectual property conventions. This negotiating mandate led to the establishment of the TRIPS Agreement during the Uruguay Round and the IPR provisions in the North American Free Trade Agreement (NAFTA). In the intervening period since the 1988 Act, the TRIPS agreement came into force and the IPR provisions of NAFTA became the template for future bilateral or regional FTAs. Thus, the focus of IPR negotiating objectives shifted from creating to strengthening the IPR trade regime.

More recent FTA negotiations have been conducted under the Trade Promotion Authority Act of 2002 (P.L. 107-210), which included two broad IPR negotiating objectives. One broad objective was to apply the existing IPR protection to digital media. The negotiating objectives contained provisions to extend IPR protection to new and emerging technologies and to methods of transmission and dissemination. The language called for standards of enforcement to keep pace with technological change and to allow rights-holders the legal and technological protections for their works over the Internet and other new media.

A second broad objective of the Trade Promotion Authority Act of 2002 was to negotiate trade agreements in terms of IPR that "reflect a standard of protection similar to that found in U.S. law."[50] This phrase opened the door to the negotiation of provisions that go beyond the level of protection provided in the TRIPS agreement. Often referred to as "TRIPS-plus" provisions, these obligations include expanding coverage to new sectors; establishing more extensive standards of protection; and reducing the flexibility options available in TRIPS. Some of the new measures also address technological innovations that have come about since the TRIPS Agreement.

In May 2007, the Bush Administration and Congress concluded a bipartisan agreement on trade policy that addressed some Members' concerns about the implications of enhanced IPR on developing countries' ability to meet public health needs.[51] In particular, congressional leadership sought to ensure that pending FTAs allowed trading partners to have enough flexibility to meet their IPR obligations and to be able to promote access to life-saving medicines, while otherwise meeting their international IPR protection and enforcement obligations. IPR language previously negotiated in the FTAs with the developing countries of Peru, Panama, and Colombia subsequently were modified to reflect the agreement. Because Korea is an industrialized country, the United States did not significantly scale-down the patent protection obligations in the U.S.-Korea FTA.

Current IPR-Related U.S. FTA Activity

In 2011, the Obama Administration may ask Congress to approve a free trade agreement (FTA) with South Korea, as well as FTAs with Colombia and Panama. These three FTAs were negotiated by the Bush Administration, but have not been implemented.

Meanwhile, the Obama Administration has entered into negotiations with participants in the Trans-Pacific Partnership (TPP) Agreement—Australia, Brunei, Chile, New Zealand, Peru, Malaysia, Singapore, and Vietnam. The objective is to build on U.S. trade ties in the region already established in FTAs with Australia, Chile, and Singapore and to provide a high-standard framework for expanded free trade in the region.[20] IPR, particularly in the area of pharmaceuticals, may prove to be a contentious issue in the negotiation of the TPP.

In addition, the Obama Administration has been conducting negotiations on the Anti-Counterfeiting Trade Agreement (ACTA), a proposed agreement being negotiated by Australia, Canada, the 27 member states of the European Union, Japan, South Korea, Mexico, Morocco, New Zealand, Singapore, Switzerland, and the United States. Negotiations on the ACTA Agreement began under the Bush Administration. The ACTA would build on the minimum standards for IPR protection and enforcement set forth by the TRIPS Agreement. It is being crafted independent of any existing international organization or agreement. The eleventh and final round of negotiations concluded on October 2, 2010, and the ACTA participants released to the public a draft text of the agreement. ACTA participants are working to resolve outstanding issues in the agreement.[52]

Central IPR Standards in U.S. FTAs

What follows is a discussion of some of the central patent and copyright standards sought in FTAs that are currently in force or have been signed by the United States (see Table 4).[53]

Patents

Patent protection is arguably the most contentious area of U.S. FTA negotiations on IPR issues. While the United States and other developed countries advocate for strong patent protections in order to promote innovation, there is concern that such stringent protections may delay developing countries' access to generic drugs and increase prices. Many of the FTAs in force include TRIPS-plus patent provisions, the most prominent of which are patent term length extensions, linkages between regulatory authority and patent status, data protection, compulsory licensing and parallel importation. The FTAs with Peru, Panama, and Colombia respond to the concerns of some Members of Congress over provisions that could restrict access to medicines in these countries and contain less ambitious standards for pharmaceutical patents, compared to previously negotiated FTAs. Pharmaceutical industry advocates express concern that this scale-down in patent protection in these FTAs may set a precedent for future FTA negotiations.[54]

Patent Term Extensions: Many FTAs include provisions for mandatory patent term length extensions beyond the TRIPS Agreement obligation of patent protection terms of twenty years from the filing date. These FTAs allow for extensions in cases of "unreasonable" delays in the issuance of patents due to the regulatory review or administrative process. Patent holders contend that such measures enhance the ability of rights-holders to recoup the costs of research and development of new products. However, there is concern that patent terms extensions may delay the entry of generic drugs into a market. In a scale-down from TRIPS-plus obligations, FTAs with Peru, Colombia, and Panama state that patent term restorations for pharmaceutical products are optional.

Patent Linkages: In general, the term "patent linkage" refers to the attachment of regulatory approval for the marketing of a drug with the status of a patent. If a patent exists, the FDA and its counterparts in other countries may not grant marketing approval for a generic version of a drug that is patented in the country without the permission of the patent holder. Patent linkage is a common provision in the trade agreements obtained by the United States. This presents a departure from TRIPS, under which generic drug manufacturers are able to apply for marketing approval without the patent owner's permission and prior to the expiration of the patent; this may reduce the time it takes for generic drugs to enter a market once the patent expires.[55] In light of developing country concerns about delays in access to generic versions of drugs, FTAs signed with Peru, Panama, and Colombia do not tie marketing approval for a generic drug with the patent status of its brand name drug.

Data Protection: In cases where the patent holders must submit undisclosed data regarding the safety or effectiveness of new pharmaceutical or agricultural products in order to market them, the TRIPS Agreement requires members to take measures to protect such data from disclosure and unfair commercial use. The TRIPS agreement does not prescribe any time period for this protection. Recent U.S. FTAs take these standards a step further, generally requiring a five-year period of marketing exclusivity for the patent holder, which typically begins from the date the product is approved in the country. Under this TRIPS-plus provision, generic drug manufacturers who want to market and distribute a generic version of a drug while the data exclusivity period is in effect must conduct their own clinical trials and submit their own findings to the national drug regulatory authority; they cannot rely on the findings submitted by the patent holder. Some critics contend that such provisions may raise the cost of manufacturing generic versions of patented drugs, as well as delay access to generic forms of drugs. The FTAs with Peru, Panama, and Colombia now include provisions that may reduce data exclusivity terms of five years by a minimum of six months in practice.[56]

Compulsory Licensing: A compulsory license is an authorization by a government for third parties (such as a company or the government itself) for the manufacture or use of a product under patent without the permission of the rights- holder. The TRIPS Agreement permits signatories to issue compulsory licenses for patented devices and provide compensation to the owner of the patent and does not limit the situations in which such licenses may be issued. The third party must have attempted to obtain permission from the patent holder, although this requirement is waived in times of national emergency or other extenuating circumstances. U.S. FTAs with Australia and Singapore limit attaining compulsory licenses only for domestic use and to situations of remedying antitrust violations or in situations of public non-commercial use, national emergency, or other cases of extreme need. Also under these FTAs, the patent holder is under no obligation to provide test data, technical know-how or other undisclosed information for the patent subject to compulsory license. The compulsory license provisions have not been included in FTAs with developing countries.

Parallel Importation: Parallel imports, also known as grey-market goods, refer to goods imported into a country without permission of the rights-holder after those goods were legitimately sold elsewhere. Parallel importation relates to the concept of territorial exhaustion of IPR, which governs the extent of IPR after the first sale. Under a national system of exhaustion practiced in the United States, IPR are exhausted domestically after the first sale, but not abroad, thus prohibiting trade in those goods without permission of the rights-holder. Under an international system, IPR are exhausted at the first sale for any destination, and such goods can be exported freely. Article 6 of the TRIPS specifically excludes issues arising from exhaustion of IPR from WTO dispute settlement, allowing each member to adopt different exhaustion regimes. Thus, TRIPS does not address the issue of parallel imports. Some developing countries contend that parallel importation is an alternative method for governments to increase access to medicines in the absence of a compulsory license.[57] Pharmaceutical companies have voiced concerns that this practice threatens their ability to engage in price differentiation between different markets. U.S. FTAs negotiated with Australia, Singapore, and Morocco disallow parallel importing of patented products. Subsequent U.S. negotiated FTAs have not included this provision, due to language included in the Science, State, Justice, and Commerce, and Related Agencies, Appropriations Act of 2006 (P.L. 109-108), which prohibited the use of such provisions.

Biodiversity and Traditional Knowledge

International trade negotiations increasingly have focused on the protection of plant and animal inventions, new plant varieties, traditional knowledge, and folklore. Some indigenous communities in developing countries and international non-governmental organizations have expressed concern about the use of patents to provide private rights for traditional knowledge and genetic material; the commercial use of such resources by entities other than the indigenous communities or countries from which such resources are derived; and the distribution of benefits from commercial use. The United States, other advanced countries, and business groups favor treating traditional knowledge and genetic material as intellectual property and protecting these resources through an IPR framework.

Article 27.3(b) of the TRIPS Agreement permits Member states to exempt "plants and animals other than microorganisms, and essentially biological processes" from patentability. TRIPS requires Members to protect plant varieties through patent protection, some other system ("sui generis"), or a combination of the two. Paragraph 19 of the Doha Declaration added another dimension to the issue by requiring the TRIPS Council to probe the relationship between the TRIPS Agreement, the UN Convention on Biological Diversity (CBD), and traditional knowledge and folklore. These issues also are being discussed in WIPO's Intergovernmental Committee on Intellectual Property and Genetic Resources, Traditional Knowledge, and Folklore (IGC).

India, Brazil, and Peru, among other countries, contend that patent applicants should be required to disclose the source of genetic materials, including plant life and traditional knowledge, before obtaining patents. The United States and the European Union have advocated for national systems in which companies are granted permission to research genetic materials and are obligated to share benefits from patents derived from those genetic products.

Some earlier U.S. FTAs have required signatories to provide protection for plants, animals, and plant varieties. The recent FTAs with Peru, Panama, and Colombia do not mandate patentability for plants and animals, but state that the countries should take efforts to expand patent coverage to these areas and to maintain this protection once it is offered. Side-letters in the three FTAs state the signatories' recognition of the importance of biodiversity and traditional knowledge and pledge the countries to work together to address these issues through the IGC.

Copyright

In the area of copyright protection, the United States has pursued certain TRIPS-plus measures in FTAs, such as extending copyright terms; including anti-circumvention provisions; and protecting rights-management information in its FTAs. The TRIPS Agreement does not mention any obligations regarding rights-management information, which is "electronic information that identifies a protected work, its author, and terms and conditions of use,"[58] perhaps due to the fact these technologies were not available at the time. In contrast, U.S.-negotiated trade agreements prohibit the removal or alteration of such information.

While patent protection has experienced policy shifts in the FTAs with Peru, Panama, and Colombia, copyright protection provisions have remained fairly consistent through the FTAs. In general, FTA signatories are obligated to provide an additional twenty years of copyright protection. This brings the minimum copyright term to seventy years from the death of the author or authorized publication, compared to fifty under the TRIPS Agreement. Responding to technological innovations not discussed in the TRIPS Agreement, many of the FTAs require trading partners to outlaw circumvention of copyrighted works. These provisions build on the U.S. Digital Millennium Copyright Act (DMCA) of 1998.[59] Also based on the DMCA, many FTAs contain provisions that regulate the liability of Internet service providers (ISPs) for copyright infringement that occurs within their networks. Under the FTAs, ISPs are provided limited immunity from copyright liability in certain kinds of infringing activities if they comply with regulations. For instance, ISPs must block access to or remove infringing materials as soon as they are aware of the infringement. Copyright holders argue that it is necessary for ISPs to assist in enforcing copyright for copyright laws to be effective. However, critics claim that these provisions impose excessive burdens on ISPs, reduce the rights of internet users, and limit the policy flexibility of FTA signatories in determining their own IPR regimes.

U.S. TRADE LAW

Special 301

Section 301 of the Trade Act of 1974 (P.L. 93-618), as amended, is the principle U.S. statute for identifying foreign trade barriers due to inadequate intellectual property protection.

The 1988 Omnibus Trade and Competitiveness Act (P.L. 100-418) strengthened section 301 by creating "Special 301" provisions, which require the USTR to conduct an annual review of foreign countries' intellectual property policies and practices. By April 30th of each year, the USTR must identify countries that do not offer "adequate and effective" protection of IPR or "fair and equitable market access to United States person that rely upon intellectual property rights."

According to an amendment to the Special 301 provisions by the Uruguay Round Agreements Act (P.L. 103-465), the USTR can identify a country as denying sufficient intellectual property protection even if the country is complying with its TRIPS commitments. These findings are submitted in the USTR's annual "Special 301" report.

Special 301 Country Lists

Within 30 days of submitting the annual National Trade Estimates of Foreign Trade Barriers report, the USTR must determine which of the identified countries are "Priority Foreign Countries."

Countries that "have the most onerous or egregious acts, policies or practices that deny intellectual property protection and limit market access to U.S. persons or firms depending on intellectual property rights protection" and "have the greatest adverse impact (actual or potential) on the relevant United States products" may be identified as "Priority Foreign Countries."

These countries may be investigated under section 301 provisions of the Trade Act of 1974. The USTR cannot identify countries as Priority Foreign Countries if they have entered into good faith negotiations or have made significant progress in improving their intellectual property protection record.[60]

If a country is named as a "Priority Foreign Country," the USTR must launch an investigation into that country's IPR practices. This investigation is conducted in a manner similar to a "Section 301" investigation; the USTR must determine a course of action within six months (9 months if a determination of complex circumstances is made).

The USTR may suspend trade concessions and impose import restrictions or duties, or enter into a binding agreement with the priority country that would eliminate the act, policy, or practice that is the subject of the action to be taken. Since the advent of the WTO and its recourse to dispute settlement, the use of the first option may lead to the initiation of dispute settlement proceedings at the WTO for member countries, rather than unilateral retaliation. For countries outside the WTO, the possibility of trade sanctions remain.

The USTR also has created several administrative categories for country identification in the Special 301 Report. "Priority Watch List" countries are those whose acts, policies, and practices warrant concern, but do not meet all of the criteria for identification as a Priority Foreign Country.

The USTR may place a country on the Priority Watch List when the country lacks proper intellectual property protection and has a market of significant U.S. interest. "Watch List" countries have intellectual property protection inadequacies that are less severe than those on the Priority Watch List, but still attract U.S. attention. Just being on one of the Special 301 lists may induce countries to improve their IPR protection.

Finally, countries identified for "Section 306" are monitored for compliance with bilateral intellectual property agreements used to resolve investigations under section 301.

Additionally, the USTR launches out-of-cycle reviews (OCRs) on countries to monitor their progress on intellectual property issues. OCRs are conducted on countries that USTR considers to require further review and may result in status changes for the following year's Special 301 report.

Special 301 Report for 2010

For the *2010 Special 301 Report*, the USTR reviewed the IPR policies and practices of 77 countries. [61]

- Despite some improvements, China and Russia remain top concerns for the Administration due to their inadequate IPR regimes.
- The USTR placed eleven countries on the Priority Watch List: Algeria, Argentina, Canada, Chile, China, India, Indonesia, Pakistan, Russia, Thailand, and Venezuela.
- The USTR placed another 29 countries were placed on the Watch List.
- No countries were designated as Priority Foreign Countries. Paraguay was identified for Section 306 monitoring.
- The USTR announced that it would conduct OCRs for the Philippines and Thailand to monitor these nations' progress on IPR and to evaluate their Special 301 status.
- Israel's status in the Special 301 Report is pending, following the conclusion of an OCR and an understanding reached by the United States and Israel on several issues in Israel's IPR regime for pharmaceutical products.[62]
- The USTR removed the Czech Republic, Hungary, and Poland from the Special 301Watch List for significant improvement in their protection and enforcement of IPR. Following an OCR on Saudi Arabia, the USTR decided to remove Saudi Arabia from the Watch List on the basis of improvements in IPR enforcement, prosecution, and transparency issues.

Country Identification Factors

Identification of countries for the "Special 301" lists is a lengthy process of information gathering and analysis based on the USTR's annual trade barriers report and consultations with a wide variety of sources, including government agencies, industry groups, other private sector representatives, Congressional leaders, and foreign governments. The Special 301 statute is the overall guideline for identifying countries for the various lists. However, placements are country-specific and, according to a USTR official, take into consideration a host of factors, several of which are mentioned in the Special 301 report.[63] These include the level and scope of the country's IPR infringement and their impact on the U.S. economy. Other considerations include the strength of the country's IPR laws and enforcement of IPR laws.

The USTR also evaluates progress made by the country in improving IPR protection and enforcement in the past year. However, even significant progress oftentimes does not change the position or inclusion of a country on the lists. For instance, the USTR may decide not to upgrade a country from the Priority Watch List to the Watch List so that it can continue monitoring the country's intellectual property practices. Also, the USTR may note significant progress made by a country but not remove the country from the Special 301 list in order to continue highlighting concerns about the country's practices and limit backsliding. Another consideration for the USTR is the sincerity of the country's commitment to multilateral and bilateral trade agreements. There is no weighting criteria for the factors or a formula to determine the placement of a country on the watch list. Furthermore, no particular threshold exists for determining when a country should be upgraded or downgraded on the list.

Some observers speculate that the Special 301 rankings are subject to external influences. The lack of a specific framework for placing countries, aside from the general directives from the Special 301 statute, has raised concerns that foreign policy considerations affect the process. For example, an IIPA representative suggested that USTR placement of countries is influenced by geopolitical reasons.[64] This source cites Russia as an example of a country with high IPR infringement that could be named as a Priority Foreign Country but is not due to unrelated foreign policy considerations. Other observers of U.S. trade policy suggest that pharmaceutical companies have a stronghold on policy direction. Oxfam International, a confederation of poverty-alleviation organizations, contends that the U.S. government's policy on patents "is still largely influenced by the narrow commercial interests of the giant pharmaceutical companies."[65] A USTR official stated that the interests of pharmaceutical companies do not override concerns by other interest groups in evaluating country placement for the Special 301 report. The official emphasized that all industry group submissions are given fair and due consideration.[66]

Section 337

Section 337 of the Tariff Act of 1930 (19 U.S.C. 1337), as amended, prohibits unfair methods of competition or other unfair acts in the importation of products into the United States. It also prohibits the importation of articles that infringe valid U.S. patents, copyrights, processes, trademarks, semiconductor products produced by infringing a protected mask work, or protected design rights. While the statute has been utilized to counter imports of products judged to be produced by unfair competition, monopolistic, or anti-competitive practices, it has become increasingly used for its IPR enforcement functions in recent years.

Under the statute, the import or sale of an infringing product is illegal only if a U.S. industry is producing an article covered by the relevant IPR exists or is in the process of being established. However, unlike other trade remedies such as antidumping or countervailing duty actions, no showing of injury due to the import is required.

The U.S. International Trade Commission (ITC) administers section 337 proceedings. USITC must investigate complaints either brought to it or ones commenced under its own initiative. An administrative law judge (ALJ) provides an initial determination (ID) to the ITC which can accept the ID or order a further review of it in whole or in part. If the ITC finds a violation, it may issue two types of remedies: exclusion orders or cease and desist orders. The ITC may issue either a limited or general exclusion order enforced by U.S. Customs.

A general exclusion order directs U.S. Customs to keep out all infringing articles regardless of the source. More commonly, a limited exclusion order is employed to exclude infringing articles from the firm subject to the ITC's investigation.

Alternatively, the ITC may enforce a cease and desist order to stop the sale of the infringing product in the United States. However, the ITC may consider several public interest criteria and decline to issue a remedy. Also, the President may disapprove a remedial order during a 60 day period for "policy reasons," which has been interpreted to mean national security reasons.[67]

During FY2009, the ITC reported a total of 85 active section 337 investigations and ancillary proceedings, comparable to the FY2008 level. Of the 85 active investigations in FY2009, 29 constituted of new section 337 investigations and 7 were new ancillary proceedings stemming from previously concluded section 337 investigations.

The section 337 investigations frequently involved advanced technology areas, such as integrated circuit, computer, telecommunications, and other electronic technologies. Over 60% of the new section 337 investigations that were active in FY2009 involved cases of alleged patent infringement. In FY2009, the ITC issued three general exclusion orders, eight limited exclusion orders, and 23 cease-and-desist orders.[68]

Generalized System of Preferences

The Generalized System of Preferences (GSP) is a program that provides preferential duty-free entry to certain products from designated developing countries. The purpose of the program is to foster economic growth in developing countries by increasing their export markets. The Trade Act of 1974 authorized the GSP for a ten-year time frame, and the program has been renewed from time to time. Most recently, in 2006, Congress extended GSP through December 31, 2010.[69]

Although the GSP is non-reciprocal, it can be used to promote stronger intellectual property protection and enforcement abroad. Under the GSP statute, the President must consider a set of mandatory criteria that a country must fulfill in order to be designated as a GSP beneficiary. Additionally, the President may evaluate a country on the basis of certain discretionary criteria, including the country's provision of IPR protection.[70]

The GSP program undergoes an annual review by the GSP Subcommittee of the Trade Policy Staff Committee (TPSC), which is headed by the USTR. As part of its evaluation, the TPSC addresses concerns about specific country practices (such as intellectual property protection) and makes recommendations to the President. For 2009, the USTR continued to evaluate IPR protection in Lebanon, Russia, and Uzbekistan on the basis of IIPA petitions for ongoing GSP reviews.[71]

U.S. Agency Functions and Funding for IPR

The United States has a complex apparatus for supporting intellectual property rights, with responsibilities cutting across many different federal government agencies. Protection activities include developing IPR policy, informing and advising Congress about IPR-related issues, participating in international trade negotiations to promote IPR, and providing IPR training and technical assistance in other countries. Enforcement activities involve the conduct of criminal investigations in the United States and abroad, interdiction of pirated and counterfeit goods, and monitoring of compliance with trade agreements. Enforcement also involves capacity-building activities to foster stronger IPR law enforcement in other countries.

It is difficult to obtain a complete picture of the magnitude of federal budget devoted to intellectual property laws. Some of these agencies perform their IPR related activities within existing budget parameters, and do not differentiate specific sums devoted to IPR-related activities. What follows is a discussion of the various IPR functions of U.S. agencies.

Department of Commerce (Commerce)

Two agencies within the Department of Commerce, the Patent and Trademark Office (PTO) and the International Trade Administration (ITA), address IPR issues.[72]

United States Patent and Trademark Office (PTO)

The PTO administers the U.S. laws pertaining to patents and trademarks. The agency processes patent and trademark applications, issues patents and registers trademarks, and circulates patent and trademark information. The PTO develops IPR protection and enforcement policy and collaborates with other agencies to develop intellectual property provisions in FTAs and other international agreements. Additionally, the PTO offers training, technical assistance, and trade capacity building programs to assist in promoting strong IPR regimes in foreign countries.[73] The PTO does not have jurisdiction over determining patent and trademark infringements; such determinations and remedies are made at the U.S. federal district court level or through the U.S. International Trade Commission's section 337 proceedings (discussed above).[74] The PTO is fully funded through fees generated from patent and trademark applications. The Consolidated Appropriations Act, 2010 (P.L. 111-117) provided $1.89 billion in budgetary authority to the PTO for FY2010, down from $2.01 billion provided for FY2009 (P.L. 111-8).

International Trade Administration (ITA)

The ITA administers many of the international trade programs of the Department of Commerce, include aspects involving IPR. The ITA monitors foreign countries' progress in implementing intellectual property agreements; reviews Generalized System of Preferences (GSP) petitions submitted by industry and coordinates the Commerce Department's response to these petitions; represents the Commerce Department at the WTO TRIPS Council; meets with trading partners to advance U.S. intellectual property interests abroad; and works with U.S. businesses and industry groups to make sure that IPR-related trade concerns are addressed.[75] For FY2010, the Consolidated Appropriations Act, 2010 (P.L. 111-117) provided the ITA with $456.20 million (including both direct appropriation and anticipated receipts from fees), up from $429.9 million provided in FY2009 (P.L. 111-8).

Department of Justice (DOJ)

The DOJ enforces criminal laws that protect IPR in the United States and internationally through the prosecution of intellectual property cases. The Civil Division's Office of Consumer Litigation specializes in intellectual property cases involving public health and safety. The Federal Bureau of Investigation (FBI) has an intellectual property enforcement program focusing on those intellectual property crimes that have the most bearing on national and economic security, such as trade secret theft, Internet piracy, and counterfeit good trafficking.[76] In addition to enforcement activities, the DOJ also works with Congress to develop laws that increase protection of IPR and provides training and technical assistance programs on IPR enforcement through its Criminal Division. The Consolidated Appropriations Act, 2010 (P.L. 111-117) provided $7.66 billion to the FBI for FY2010, up from the FY2009 level of $7.07 billion (P.L. 111-8).

Department of Homeland Security (DHS)

The DHS, through its Customs and Border Protection (CPB) unit and Immigration and Customs Enforcement (ICE) unit, conducts intellectual property rights enforcement activities. Neither DHS unit has a line item for IPR enforcement. The ICE and the FBI jointly run the National Intellectual Property Rights Coordination Center that coordinates U.S. Government domestic and international law enforcement activities.[77]

Customs and Border Protection (CBP)

Taking the lead in day-to-day IPR enforcement activities at the U.S. border, the CBP is responsible for detecting and seizing counterfeit and pirated goods entering the United States and determining penalties for infringement.[78] CBP has the authority to determine whether or not imports infringe federally registered trademarks and copyrights and to detain or seize such infringing goods. Owners of copyrights and trademarks are able to record information about their rights in the CBP's electronic IPR database. As noted earlier, in contrast to trademarks and copyrights, CBP does not have the jurisdiction to make determinations about patent infringements. However, it is able to block imports determined by the ITC to infringe a U.S. patent by a Section 337 investigation.[79] For FY2010, the Consolidated Appropriations Act, 2010 provided $8.07 billion in budget authority to CBP, up from $7.60 billion in FY2009.

Immigration and Customs Enforcement (ICE)

ICE is charged with investigating violations of U.S. law that are connected with U.S. borders. ICE "identifies, investigates, apprehends, and removes" international criminal groups and other criminals. ICE conducts inquiries into the importation and distribution of counterfeit goods. ICE activities are closely linked with those of CBP. For instance, when CBP identifies and seizes counterfeit goods, the issue is referred to ICE for criminal investigation. Likewise, information obtained from ICE activities that is relevant to identifying and apprehending counterfeit shipments is provided to CBP.[80] For FY2010, the Consolidated Appropriations Act, 2010 provided $5.34 billion in budget authority to ICE, up from $4.93 billion in FY2009.

Food and Drug Administration (FDA)

The FDA, which is an agency of the Department of Health and Human Services (DHHS), is responsible for protecting public health by ensuring the safety and effectiveness of medicines, food, and other products. As part of its activities, the FDA works to protect consumers against counterfeit medicines. To combat the entry of foreign counterfeit drugs into the U.S. drug supply, the FDA works in conjunction with the CBP to conduct border inspections of FDA-regulated products. The FDA also engages in foreign inspections to ensure that foreign manufacturers meet FDA quality and labeling requirements. Funding for preventing counterfeits from entering the United States is part of overall FDA import safety efforts.[81] The enacted FY2010 Agricultural appropriation, P.L. 111-80, contained $2.36 billion in budget authority for the FDA, up from $2.05 billion for FY2009.

Copyright Office

The Copyright Office of the Library of Congress administers U.S. copyright law by registering claims to copyright and related documents, including "assignments or transfers of rights" and maintains information on registrations, recordings, compulsory licenses, and other copyright-related actions. Additionally, the Copyright Office provides legal and technical expertise on national and international copyright issues to the U.S. government. The Copyright Office also works with other federal agencies to provide assistance and advice in negotiations for international intellectual property agreements, as well as technical assistance to foreign countries crafting their own copyright laws.[82] The enacted FY2010 Legislative appropriation, P.L. 111-68, contained $20.9 million in new budget authority (not including authority to spend $33.3 million in receipts) for the Copyright

Office, up from $18.3 million in new budget authority in FY2009 (not including authority to spend $33.3 million receipts).

Copyright Office appropriations also specify funding for IPR-related activities in developing countries. As in previous years, the FY2010 appropriations act provided $100,000 to the International Copyright Institute for such activities.

Department of State (State)

State represents U.S. views in both bilateral and multilateral arenas. State works to build international consensus for intellectual property rights enforcement. Information from State's foreign postings informs the USTR Special 301 review. In particular, the Bureau of International Narcotics Control and Law Enforcement (INCLE) works to combat intellectual property piracy, while the Bureau of Energy, Economics and Business Affairs supports stronger international IPR standards to fight global piracy and counterfeiting.[83] As in previous years, the Consolidated Appropriations Act, 2010 (P.L. 111-117) provided $5 million from the INCLE Account for combating copyright piracy (Section 688).

U.S. Agency for International Development (AID)

AID funds training and technical assistance to improve the compliance with the TRIPS Agreement and bilateral trade agreements with the United States. Funding for these projects generally have been undertaken by regional or country missions; there is no separate budgetary line item for IPR enforcement and training.[84]

United States Trade Representative (USTR)

The USTR is the lead trade agency of the United States government. Through its annual Special 301 report, USTR is charged with monitoring the adequacy and effectiveness of IPR protection of our trading partners as well as their compliance with bilateral and multilateral trade agreements, to identify countries not in compliance with such agreements, and to negotiate with those countries better compliance. USTR also advances greater protection and enforcement of IPR in its negotiations of U.S. free trade agreements. Additionally, USTR works to implement the Administration's STOP! Initiative, which draws together the major federal government agencies, private sector groups, and trading partners to take targeted action in fighting piracy and counterfeiting.[85] The FY2010 funding level for USTR was $47.83 million, under the Consolidated Appropriations Act, 2010 (P.L. 111-117), up from the FY2009 funding level of $47.3 million. Explanatory language in H.Rept. 111-149, accompanying P.L. 111-117, encouraged the USTR to continue prioritizing IPR issues with China, Russia, and Canada in bilateral and multilateral trade negotiations.

United States International Trade Commission (ITC)

The ITC is a quasi-judicial federal government agency responsible for investigating and arbitrating complaints of unfair trade practices. The ITC adjudicates allegations of imported products that infringe U.S. patents, trademarks, and copyrights through its section 337 proceedings (see above). The primary remedy employed by the ITC is to order the CBP to stop imports from entering the border. Additionally, the ITC may issue "cease and desist" orders against individuals determined to be IPR violators. Damages for IPR infringement cannot be received through ITC court proceedings; rights-holders seeking damages must file action with the U.S. federal district court.[86] For FY2010, the Consolidated Appropriations Act, 2010 (P.L. 111-117) provided the ITC with $81.86 million in budget authority, up from the FY2009 level of $75.1 million.

Intellectual Property Enforcement Coordinator (IPEC)

In October 2008, Congress created an Intellectual Property Enforcement Coordinator (IPEC), through the Prioritizing Resources and Organization for Intellectual Property Act of 2008 (P.L. 110-403). The IPEC, located in the Executive Office of the President – specifically in the Office of Management and Budget – and subject to Senate confirmation, is charged with coordinating U.S. government agency IPR enforcement actions and with providing assistance to the USTR in conducting trade negotiations relating to IPR enforcement abroad.[87]

Under P.L. 110-403, the IPEC is to chair a Advisory Committee composed of representatives from the Office of Management and Budget, the Departments of Justice, Commerce, State, Homeland Security, Agriculture, the Food and Drug Administration, the Agency for International Development, and the Register of Copyrights.

The IPEC, assisted by its Advisory Committee, also is charged with developing a "Joint Strategic Plan" for combating counterfeiting and piracy. Legislation requires the Joint Strategic Plan to include in its objectives: reducing counterfeiting and infringing goods in the domestic and international supply chain, identifying and addressing barriers to effective enforcement domestically, ensuring that information is shared among the relevant departments and agencies, eliminating domestic and international counterfeiting and infringement networks, strengthening the capacity of foreign countries to protect and enforce IPR, and cooperating with other countries to establish international standards and policies to enforce IPR.

In 2009, President Obama nominated and the Senate confirmed Victoria A. Espinel as the IPEC. In June 2010, IPEC Espinel released the *2010 Joint Strategic Plan on Intellectual Property Enforcement*. The plan identifies 33 "enforcement strategy action items," categorized under six major areas of focus: (1) leading by example; (2) increasing transparency; (3) ensuring efficiency and coordination; (4) enforcing U.S. rights internationally; (5) securing the supply chain; and (6) building a data-driven government.[88]

The House Conference Report (H.Rept. 111-366 to H.R. 3288) provided for a transfer of $176,000 from the White House to the OMB, to reflect the Administration's decision to locate the new IPEC in the OMB rather than the White House.

ISSUES FOR CONGRESS

The role of Congress in addressing IPR and trade-related issues stems from the U.S. Constitution, which provides Congress with the power "To promote the Progress of Science and useful Arts, by securing for limited Times to Authors and Inventors the exclusive Right to their respective Writings and Discoveries" and "To regulate Commerce with foreign Nations."[89] While the U.S. Constitution provides Congress with the power to regulate international trade, the authority to negotiate trade agreements has been delegated periodically to the President. However, congressional action is needed to bring the trade agreements into force.

From a policy perspective, congressional consideration of IPR may take place in the context of the National Export Initiative (NEI), an Obama Administration plan introduced in early 2010 to double U.S. exports in five years to create two million U.S. jobs.[90] A report submitted by the President's Export Promotion Cabinet on implementing the NEI discussed the relationship between strengthening IPR regimes internationally and promoting U.S. exports.[91] In December 2010, the President's Export Council sent a letter to President Obama highlighting the importance of addressing inadequate protection and enforcement of IPR as a means for "boosting exports and foreign sales, and promoting the sustained growth of well compensated U.S. jobs." The President's Export Council expressed support for making "efforts to combat weak and ineffective intellectual property regimes abroad an integral and essential part of the National Export Initiative."[92]

U.S. Efforts to Promote IPR through Trade Policy

Since the inclusion of IPR provisions in the TRIPS Agreement, there has been an ongoing debate about the appropriateness of including IPR as a component of U.S. trade policy. Some argue that IPR, which grant legal temporary monopolies to rights-holders for their creations, are actually barriers to trade and have no place in trade liberalization negotiations. Others contend that IPR promote trade through innovation, economic growth, and technology transfer from advanced to developing countries.

In addition to this broader discussion about the role of IPR in trade policy, concerns have been voiced about the trade policy channels used by the United States to promote international IPR protection and enforcement. Some question the appropriateness of using regional and bilateral FTAs for this pursuing stronger IPR, contending that such actions take away from the effectiveness of multilateral IPR promotion efforts. Periodically, Members of Congress have expressed concern over U.S attempts to expand the IPR obligations of foreign countries through trade agreements. In 2002, the Trade Promotion Authority (TPA) legislation was amended to state that the United States recognized the Doha Declaration on the TRIPS Agreement and Public Health in the context of negotiating FTAs. In the 110th Congress, congressional leaders and the Bush Administration agreed to modifications of the patent provisions in the Peru FTA as a result of the May 2007 bipartisan trade agreement. Still, a Government Accountability Office (GAO) report suggests that USTR should offer clearer policy guidance to align FTA negotiating activities with the WTO Doha Declaration.[93] Additionally, there is concern by some that the ratchet of IPR commitments pursued through regional and bilateral FTAs may be too stringent for developing countries and may limit innovation and creativity by stifling the exchange of ideas.

Further expansion of IPR provisions may be affected by the language of any future TPA. In discussions about renewal of TPA, Congress may choose to consider possible reiteration or expansion on its IPR goals related to global health from the 2002 TPA. Congress also may choose to consider whether or not to follow the template provided by the Peru, Panama, and Colombia FTAs in future trade negotiations, such as negotiations on the Trans-Pacific Partnership Agreement.

Effectiveness of the U.S. IPR Organizational Structure

There are concerns on the part of some lawmakers about whether or not the present U.S. IPR organizational structure is doing enough to enforce foreign countries' IPR obligations, as well as concerns about whether or not the structure is capable of doing more.

Prior to the establishment of the Intellectual Property Enforcement Coordinator (IPEC), the U.S. IPR organizational structure was coordinated by the National Intellectual Property Law Enforcement Coordinating Council (NIPLECC). Some Members of Congress were critical of NIPLECC's organizational response to international IPR protection and enforcement. Recent GAO testimony pointed out some of the problems associated with NIPLECC, including an absence of mission, dearth of activities, and poor image among businesses.[94] The FY2007 CJS Committee Report (S.Rept. 109-280) expressed concern about the lack of information on NIPLECC's progress and evidence of success.

In the 110th Congress, several bills were introduced to repeal and replace NIPLECC. The Prioritizing Resources and Organization for Intellectual Property Act of 2008 (P.L. 110-403) was signed by President Bush on October 13, 2008. The act, among other provisions, replaced NIPLECC with an IPEC. Elevating the IPEC's profile by placing it in the EOP, some lawmakers hope, will promote more effective coordination of the organizational structure's response to international IPR protection and enforcement issues. Others argue that the effectiveness of the IPEC will depend on what resources are allocated for its activities.

While protection and enforcement of IPR is a stated trade policy priority for the United States, it is difficult to get a sense of the magnitude of funding and resources devoted toward IPR support. Some agencies do not have a separate budgetary line item for IPR-related activities, and Congress does not always designate specific funds for IPR activities in its appropriations for agencies.[95] Additionally, there is limited information on the economic and other impacts of piracy and counterfeiting on the United States. For example, in its Special 301 Report, USTR uses industry figures that are not independently confirmed. This may complicate the ability of lawmakers to weigh the threat of IPR infringement against the federal resources available for IPR and other government priorities.

Legislation was introduced in the 111th Congress to advance U.S. IPR protection and enforcement efforts as part of U.S. trade policy. The Trade Enforcement Act of 2009 (H.R. 496, S. 1466) would have created new IPR coordinator positions in the Department of the Treasury and the Department of Homeland Security's Customs and Border Protection (CBP) and Immigration and Customs Enforcement (ICE) agencies; increased IPR resources, staff, funding, and training for CBP and ICE; required the development of a strategy for IPR enforcement; and created a Advisory Committee on Import Safety and Intellectual Property Rights Enforcement, among other provisions.

Some lawmakers also support increasing the priority that IPR is given as part of U.S. foreign policy. Foreign relations appropriations legislation for FY2010 (H.R. 2410, H.R. 2475) would have required the Secretary of State to appoint ten intellectual property attachés to serve in U.S. missions overseas and would have directed the Secretary of State to consider assigning such attachés to missions in countries which have been identified under Section 182 of the Trade Act of 1974.

Some may support efforts to promote IPR as a part of U.S. trade, foreign, or other forms of policy. Others may raise concerns about how the promotion of IPR may affect U.S. efforts to advance other policy goals. In addition, while some support efforts to increase resources dedicated to IPR protection and enforcement, others question what implications such increased resources might have for U.S. coordination of IPR protection and enforcement activities.

End Notes

[1] Uniform Trade Secret Act, §1(4).

[2] For information on geographical indications and international trade negotiations, see CRS Report RS21569, *Geographical Indications and WTO Negotiations*, by Charles E. Hanrahan.

[3] Counterfeit goods should be distinguished from generic goods, i.e., in the case of generic forms of pharmaceutical medicines.

[4] WIPO, "Global Economic Slowdown Impacts 2008 International Patent Filings," press release, January 27, 2009, http://www.wipo.int/pressroom/en/articles/2009/article_0002.html.

[5] WIPO, "International Patent Filings Recover in 2010," press release, February 9, 2011, http://www.wipo.int/pressroom/en/articles/2011/article_0004.html.

[6] Stephen E. Siwek, "Engines of Growth: Economic Contributions of the US Intellectual Property Industries," commissioned by NBC Universal, 2005, p. 2.

[7] Arti Rai, Stuart Graham, and Mark Doms, *Patent Reform: Unleashing Innovation, Promoting Economic Growth & Producing High-Paying Jobs*, Department of Commerce, A White Paper from the U.S. Department of Commerce, April 13, 2010.

[8] Nam D. Pham, *The Impact of Innovation and the Role of Intellectual Property Rights on U.S. Productivity, Competitiveness, Jobs, Wages, and Exports*, NDP Consulting Group, April 2010.

[9] Stephen E. Siwek, *Copyright Industries in the U.S. Economy: The 2003-2007 Report*, prepared for the International Intellectual Property Alliance (IIPA), http://www.iipa.com, pp. 3-7.

[10] PhRMA, *Pharmaceutical Industry Profile 2010*, Washington, D.C., March 2010, http://www.phrma.org, p. 50.

[11] Amanda Horan, Christopher Johnson, and Heather Sykes, *Foreign Infringement of Intellectual Property Rights: Implications for Selected U.S. Industries*, U.S. International Trade Commission, Office of Industries Working Paper, October 2005, p. 4.
[12] Jennifer Koncz and Anne Flatness, "U.S. International Services," *Survey of Current Business*, U.S. Bureau of Economic Analysis (BEA), October 2010. This measure of cross-border trade in royalties and license fees by U.S. companies include transactions with both affiliated and unaffiliated foreign companies.
[13] Thomas Rogers and Andrew Zamosszegi, *Fair Use in the U.S. Economy:Economic Contribution of Industries Relying on Fair Use*, Prepared for the Computer & Communications Industry Association (CCIA), 2010. CCIA, "Fair Use Doctrine Vital for All of Us," press release, November 18, 2009, http://www.ccianet.org/index.asp?sid= 5&artid=125&evtflg=False. See also CRS Report RL33631, *Copyright Licensing in Music Distribution, Reproduction, and Public Performance*, by Brian T. Yeh.
[14] Ibid., p. 2.
[15] OECD, *The Economic Impact of Counterfeiting and Piracy*, 2007, p. 4, http://www.oecd.org/ dataoecd/ 13/12/38707619.pdf.
[16] Organization for Economic Cooperation and Development (OECD), *The Economic Impact of Counterfeiting and Piracy*, 2007. OECD, *Magnitude of Counterfeiting and Piracy of Tangible Products: An Update*, November 2009.
[17] U.S CBP and U.S. ICE, *Intellectual Property Rights Seizure Statistics: FY2009*, October 2010.
[18] Ibid.
[19] Ibid.
[20] There may be limitations on data estimating the impact of counterfeiting and piracy on the U.S. economy. Some critics point out that many of the estimates for losses associated with IPR infringement are generated by industry groups that may have self-interested motivations and hence, the negative effects may be exaggerated.
[21] The IIPA member associations are: Association of American Publishers (AAP), Business Software Alliance (BSA), Entertainment Software Alliance (ESA), Independent Film & Television Alliance (I.F.T.A.), Motion Picture Association of America (MPAA), National Music Publishers' Association (NMPA), and Recording Industry Association of America (RIAA).
[22] IIPA, *IIPA 2009 "Special 301" USTR Decisions*, issued June 30, 2009.
[23] PhRMA's calculations of damage due to violations of data exclusivity are based on a five-year data protection period; any sales not made by the patent holder within the data exclusivity period were regarded as data exclusivity damages. For damages from patent protection violations, PhRMA used a ten-year patent protection period and considered any sales not made by the patent holder within that period to be damages. For some countries, PhRMA did not report damages because data was not available at the time.
[24] Economic losses in PhRMA's annual Special 301 submission are not reported on calendar-year basis because fourth quarter economic data is not available at the time the report is issued.
[25] PhRMA, *Special 301 Submission for 2007.* See Appendix: Damage Estimate Methodology, p. v.
[26] PhRMA, *Special 301 Submission 2008.*
[27] U.S. Government Accountability Office, *Intellectual Property: Observations on Efforts to Quantify the Economic Effects of Counterfeit and Pirated Goods*, GAO-10-423, April 2010.
[28] Monika Ermert, "Speculation Persists on ACTA as First Official Meeting Concludes," *Intellectual Property Watch*, June 5, 2008. Robin Gross, *IP Justice White Paper on the Proposed Anti-Counterfeiting Trade Agreement (ACTA)*, IP Justice, March 25, 2008, p. 6.
[29] Robert G. Picard, "A Note on Economic Losses Due to Theft, Infringement, and Piracy of Protected Works," *Journal of Media Economics*, 17(3), 207-217, 2004.
[30] U.S. Government Accountability Office, *Intellectual Property: Observations on Efforts to Quantify the Economic Effects of Counterfeit and Pirated Goods*, GAO-10-423, April 2010.

[31] IIPA, "IIPA's 2009 Special 301 Report on Copyright Protection and Enforcement," submitted to the U.S. Trade Representative by IIPA on February 17, 2009, http://www.iipa.com.

[32] Envisional, *Technical Report: An Estimate of Infringing Use of the Internet*, January 2011.

[33] USTR, "United States Files WTO Cases Against China Over Deficiencies in China's Intellectual Property Rights Laws and Market Access Barriers to Copyright-Based Industries," press release, April 9, 2007, http://www.ustr.gov. See also CRS Report RL33536, *China-U.S. Trade Issues*, by Wayne M. Morrison.

[34] WTO, "WTO issues panel report on U.S.-China dispute over intellectual property rights," press release, January 26, 2009. USTR, "United States Wins WTO Dispute Over Deficiencies in China's Intellectual Property Rights Law," press release, January 26, 2009. Daniel Pruzin, "WTO Publishes Final Ruling in U.S. Complaint Against Chinese IPR Enforcement Measures," *International Trade Daily*, January 27, 2009.

[35] For more information on U.S-China IPR and other trade issues, see CRS Report RL33536, *China-U.S. Trade Issues*, by Wayne M. Morrison.

[36] Pascal Lamy, "Trade-Related Aspects of Intellectual Property Rights - Ten Years Later," *Journal of World Trade,* October 2004, p. 925.

[37] "Extension of the Transition Period under Article 66.1 of the TRIPS Agreement for Least-Developed Country Members for Certain Obligations with Respect to Pharmaceutical Products," WTO Document IP/C/25, July 1, 2002.

[38] See CRS Report R40607, *Intellectual Property Rights and Access to Medicines: International Trade Issues*, by Shayerah Ilias.

[39] Declaration on the TRIPS Agreement and Public Health, (WT/MIN(01)/DEC/2), November 14, 2001, available at http://www.wto.org/english/thewto_e/minist_e/min01_e/ mindecl_ trips_e.htm.

[40] "Implementation of Paragraph 6 of the Doha Declaration on the TRIPS Agreement and Public Health," IP/C/W/405, August 30, 2003, and accompanying Chairman's statement, available at http://www.wto.org/english/news_e/pres03_e/ pr350_e.htm.

[41] "Members accepting amendment of the TRIPS Agreement," http://www.wto.org/english/tratop_e/trips_e/ amendment_e.htm.

[42] "Canada Issues Compulsory License For HIV/AIDS Drug Export To Rwanda, In First Test Of WTO Procedure," *Bridges Weekly Trade News Digest*, September 26, 2007, http://www.ictsd.org/weekly/07-09-26/story2.htm.

[43] OECD, *The Impact of Trade-Related Intellectual Property Rights on Trade and Foreign Direct Investment in Developing Countries*, May 28, 2003, p. 21, http://www.oecd.org.

[44] Keith E. Maskus, *Intellectual Property Rights in the Global Economy*, Institute for International Economics (IIE), Washington, D.C., August 2000.

[45] Commission on Intellectual Property Rights (CIPR), *Integrating Intellectual Property Rights and Development Policy*, September 2002.

[46] These WCT and WPPT frequently are referred to as the WIPO Internet Treaties.

[47] Monika Ermert, "G8 Governments Want ACTA Finalised This Year, SPLT Talks Accelerated ," *Intellectual Property Watch*, July 9, 2008.

[48] "Agreement between the World Intellectual Property Organization and the World Trade Organization," http://www.wto.org/english/tratop_e/trips_e/wtowip_e.htm.

[49] WIPO, *Intellectual Property Handbook*, http://www.wipo.org, p. 359.

[50] P.L. 107-210, Sec. 2102(b)(4).

[51] The May 10, 2007 bipartisan trade agreement is available online at: http://www.ustr.gov/assets/Document_Library/Fact_Sheets/2007/asset_upload_file127_113 19.pdf.

[52] For more information, see CRS Report R41107, *The Proposed Anti-Counterfeiting Trade Agreement: Background and Key Issues*, by Shayerah Ilias.

[53] For a more detailed discussion of the differences between the TRIPS Agreement and regional FTAs that are in force, see CRS Report RL33205, *Intellectual Property and the Free Trade Agreements: Innovation Policy Issues*, by John R. Thomas.

[54] "Brand-Name Industry Alarmed at IPR Precedent of FTA Template," *Inside US Trade*, May 18, 2007.

[55] While TRIPS does not directly speak to the rights of generic drug manufacturers in obtaining marketing approval for a generic drug before the expiration of the patented drug, Article 30 of TRIPS permits exceptions of patent rights for activities such as "research, prior user rights, and pre-expiration testing."

[56] For example, under the Peru FTA, if a company files to market a new drug in Peru after making an initial filing in another country, such as the United States, and Peru approves the drug within six months of the filing, the data exclusivity period begins at the time the drug was approved in the country of the initial filing, not Peru.

[57] U.S. Government Accountability Office, *U.S. Trade Policy Guidance on WTO Declaration on Access to Medicines May Need Clarification*, GAO-97-1198, September 2007, p. 19.

[58] CRS Report RL33205, *Intellectual Property and the Free Trade Agreements: Innovation Policy Issues*, by John R. Thomas.

[59] The DMCA (P.L. 105-304) prohibits disabling technological protection measures designed to protect copyright works through activities such as descrambling or decrypting copyrighted workers.

[60] For the Special 301 provisions, see 19 U.S.C. §2242; Trade Act of 1974, as amended, (P.L. 93-618), §182.

[61] The USTR *Special 301 Report 2010* is available on the USTR website, http://www.ustr.gov.

[62] Letter from Ambassador Ron Kirk, U.S. Trade Representative, to The Honorable Benyamin Ben-Eliezer, Ministry of Industry, Trade and Labor, February 18, 2010.

[63] Conversation with USTR official, July 2006.

[64] Telephone conversation with PhRMA representative, July 2006.

[65] Oxfam International, *US bullying on drug patents: One year after Doha*, Briefing Paper 33.

[66] Telephone conversation with USTR official, July 2006.

[67] For more information on the Section 337 investigation and enforcement process, see CRS Report RS22880, *Intellectual Property Rights Protection and Enforcement: Section 337 of the Tariff Act of 1930*, by Shayerah Ilias.

[68] U.S. International Trade Commission, *Year in Review: Fiscal Year 2009*, USITC Publication 4167, p. 14.

[69] For a more thorough discussion of GSP, see CRS Report RL33663, *Generalized System of Preferences: Background and Renewal Debate*, by Vivian C. Jones.

[70] 91 USC 2462(b)(2)

[71] USTR, *GSP: 2009 Annual Review*, http://www.ustr.gov.

[72] General information about the Department of Commerce is available at http://www.doc.gov.

[73] NIPLECC, *Report to the President and Congress on Coordination of Intellectual Property Enforcement and Protection*, January 2008, p. 21.

[74] Conversation with PTO official, November 26, 2007.

[75] NIPLECC, *Report to the President and Congress on Coordination of Intellectual Property Enforcement and Protection*, January 2008, p. 21.

[76] DOJ, *Progress Report of the Department of Justice's Task Force on Intellectual Property*, June 2006, http://www.usdoj.gov/opa/documents/ipreport61906.pdf, pp. 17-24.

[77] Information about the DHS is available at http://www.dhs.gov.

[78] Certain customs-related IPR policy-making resides within the Treasury.

[79] NIPLECC, *Report to the President and Congress on Coordination of Intellectual Property Enforcement and Protection*, January 2008, pp. 15-16. Additional information about CBP is available at http://www.cbp.gov.

[80] Ibid. Also refer to the ICE website, http://www.ice.gov.

[81] Conversation with FDA official, November 26, 2007. Additional information is available on the FDA website, http://www.fda.gov.

[82] NIPLECC, *Report to the President and Congress on Coordination of Intellectual Property Enforcement and Protection*, January 2008, p. 18. Also see Copyright Office website, http://www.copyright.gov.

[83] NIPLECC, *Report to the President and Congress on Coordination of Intellectual Property Enforcement and Protection*, January 2008, pp. 17-18. Additional information about the State Department is available at http://www.state.gov.

[84] Trade Capacity Database and general AID information is accessible at http://www.usaid.gov.

[85] NIPLECC, *Report to the President and Congress on Coordination of Intellectual Property Enforcement and Protection*, January 2008, p. 25. Also see USTR website, http://www.ustr.gov.

[86] U.S. ITC website, http://www.usitc.gov.

[87] In creating the IPEC, P.L. 110-403 repealed the authorities creating the National Intellectual Property Law Enforcement Coordination Council (NIPLECC). Established by Congress in 1999, NIPLECC coordinated U.S. activities to protect and enforce IPR domestically and abroad, drawing together the major federal agencies the help to enforce IPR. The Copyright Office participated in the Council in an advisory role. The U.S. Coordinator for International Intellectual Property Enforcement headed NIPLECC's interagency coordination efforts. NIPLECC, *Report to the President and Congress on Coordination of Intellectual Property Enforcement and Protection*, January 2008, pp. 3-4.

[88] Executive Office of the President, *2010 Joint Strategic Plan on Intellectual Property Enforcement*, June 2010, http://www.whitehouse.gov/sites/default/files/omb/assets/intellectualproperty/intellectualproperty_strategic_plan.pdf.

[89] *U.S. Constitution*, Article 1, Section 8.

[90] "National Export Initiative," Executive Order 13534 of March 11, 2010, 75 *Federal Registrar* 12433, March 16, 2010.

[91] Export Promotion Cabinet, *Report to the President on the National Export Initiative: The Export Promotion Cabinet's Plan for Doubling U.S. Exports in Five Years*, Washington, D.C., September 2010. The President's Export Promotion Cabinet is a high-level body comprised of the Secretaries or Directors of key federal agencies involved in U.S. export promotion efforts.

[92] Letter from Jim McNerney, The President's Export Council, to President Obama, December 9, 2010, http://trade.gov/pec/docs/PEC_IPR_Letter_120910.pdf. The President's Export Council is a private sector national advisory committee on international trade.

[93] U.S. Government Accountability Office, *U.S. Trade Policy Guidance on WTO Declaration on Access to Medicines May Need Clarification,* GAO-07-1198, September 2007.

[94] U.S. Government Accountability Office, *National Enforcement Strategy Needs Stronger Leadership and More Accountability*, GAO-07-710T, April 12, 2007, pp. 8-10.

[95] Liza Porteus Viana, "US Fiscal 2009 Proposed Budget Shows IP Enforcement a Priority," *Intellectual Property Watch*, February 6, 2008.

INDEX

A

B

C

J

K

L

M

N

O

P

R

S

T

U

V

W